Introductory Astronomy Exercises

SECOND EDITION

Dale C. Ferguson
Baldwin-Wallace College

BROOKS/COLE

THOMSON LEARNING

Australia • Canada • Mexico • Singapore • Spain • United Kingdom • United States

BROOKS/COLE

THOMSON LEARNING

Sponsoring Editor: Jennifer Huber	*Proofreader:* Kristen Cassereau
Project Development Editor: Marie Carigma-Sambilay	*Permissions Editor:* Sue Ewing
Editorial Associate: Samuel Subity	*Cover Design:* Denise Davidson
Production Coordinator: Stephanie Andersen	*Cover Photos:* Neelon Crawford
Production Service: Scratchgravel Publishing Services	*Print Buyer:* Micky Lawler
	Printing and Binding: Malloy Litho, Inc.

For more information about this or any other Brooks/Cole products, contact:
BROOKS/COLE
511 Forest Lodge Road
Pacific Grove, CA 93950 USA
www.brookscole.com
1-800-423-0563 (Thomson Learning Academic Resource Center)

For permission to use material from this work, contact us by
www.thomsonrights.com
fax: 1-800-730-2215
phone: 1-800-730-2214

Printed in the United States of America

10   9   8   7   6   5   4   3   2   1

Library of Congress Cataloging-in-Publication Data

Ferguson, Dale C.
    Introductory astronomy exercises / Dale C. Ferguson.—2nd ed.
        p. cm.
    ISBN 0-534-37977-X
    1. Astronomy—Laboratory manuals. 2. Astronomy—Problems, exercises, etc. I Title.
QB62.7 .F47 2001
522'.078—dc21   00-051927

Contents

Preface   vii

Extra Equipment and Materials Needed for the Exercises   x

Introduction to the Science of Astronomy

Exercise 1   Introduction to Astronomical Telescopes   1

Exercise 2   Constellations and the Celestial Sphere   7

Exercise 3   Introduction to Experimental Measurements   13

The Solar System

Exercise 4   Interplanetary Travel*   19

Exercise 5   The Seasons*   25

Exercise 6   The Temperature of the Earth   31

Exercise 7   Astronomical Systems of Time   39

Exercise 8   Observing with the Telescope, Part 1:
Locating Celestial Objects*   45

Should be done in a relatively permanent lab room.

Exercise 9   Optics in Astronomy* 55

Exercise 10   Observing with the Telescope, Part II:
The Limitations of the Telescope   61

Exercise 11   The Moon   67

Exercise 12   Observing with the Telescope, Part III:
Visual Observations of the Moon   97

Exercise 13   The Planets, Part I: Analysis of Observations   101

Exercise 14   The Planets, Part II: Observations with the Telescope   125

Exercise 15   Observing with the Telescope, Part IV:
Visual Observations of the Sun   129

Stellar Astronomy

Exercise 16   Measurement of Astronomical Distances   137

Exercise 17   Kepler's Third Law and Masses in Astronomy   153

Exercise 18   Photoelectric Photometry   163

Exercise 19   Spectroscopy in Astronomy* 175

Exercise 20   Spectral Classification   181

Exercise 21   The Hertzsprung–Russell Diagram   205

Exercise 22   Telescopic Observing with Equatorial Mounting, Clock
Drive, Setting Circles, and Slow Motion Controls   229

Exercise 23   Pulsars   233

Exercise 24   Galactic Spiral Structure   245

Exercise 25   Astronomical Image Processing   255

Exercise 26   CCD Photography at the Telescope   263

Exercise 27   Classification of Galaxies   269

Exercise 28   Radial Velocities and the Hubble Law   295

Appendix A   Fall Observing List   305

Appendix B   Finding List for October Celestial Objects   309

Appendix C   Spring Observing List   311

Appendix D   Field Observing in March 315

*Should be done in a relatively permanent lab room.

Preface

The laboratory exercises in this manual have, by and large, stood the test of time. They have been taught at the University of Arizona, Louisiana State University, New York University, Southeast Missouri State University, and Baldwin-Wallace College in Ohio within the past 15 to 20 years. Nevertheless, they are remarkably up to date, with a few necessary modifications just recently made.

All of the exercises may be done within a lab class time of two or three hours, and all of them illustrate the methods and concepts of astronomy. They require a variety of mostly inexpensive equipment available at most colleges offering astronomy courses. The labs herein are not completely homogeneous; they have had various origins and authors, most of which have become anonymous through the passage of time. However, I feel confident that they will be useful and instructive to beginning astronomy students.

The first three exercises are basic to any introductory astronomy lab course. The first of these is called An Introduction to Astronomical Telescopes. Students should learn how telescopes work and how to use small telescopes as soon as possible, because telescopes are the tools of astronomers and because their use will open up a fascinating new world to the beginner.

The second exercise is Constellations and the Celestial Sphere. The ability to find one's way around the sky is of great interest to students; it is also necessary in order for them to be able to complete the Observing List, which may be incorporated as part of the course. The Observing List, designed to be completed during about the last half hour to hour of each class session, motivates students to become familiar with the night sky and to learn the use of small telescopes.

The third fundamental exercise is on the use of numbers in science and is called Introduction to Experimental Measurements. This exercise will save students untold time and worry throughout the course; it tells them how many decimal places to carry in their calculations, how to use scientific notation, and how to use the calculator for computation. These skills, if mastered in the first weeks of the course, will be especially useful to students who have an insufficient or only dimly remembered mathematical background.

Included with the laboratory exercises (as Appendices A and C) are Fall and Spring Observing Lists. These contain many of the interesting objects in the skies and may help students become familiar with the constellations, stars, and other astronomical features visible with the naked eye or a small telescope. They suggest a number of objects of each type to be found during a one-term course. Appendices B and D are suggested "observing night" exercises, which may be done on field trips or as dedicated observing labs. Although these are keyed to specific seasons, for the most part the other exercises may be done at any season with little or no modification. When necessary, modifications will be indicated in the lists of equipment and materials needed.

In general, the first set of exercises is concerned with the solar system and the second with stars and objects beyond the solar system. There are more exercises for each set (counting the fundamental introductory exercises) than there are usable class periods in a quarter, and more in the entire manual than there are classes in a semester. Therefore, in a one-quarter or one-semester course, many of the exercises must be skipped. This is a matter for the individual instructor to decide.

All exercises are marked Indoor or Outdoor on their starting pages. If inclement weather prohibits using an outdoor exercise, the next indoor exercise may be substituted on short notice and will usually turn out all right.

In addition, most of the exercises are readily changed according to the desires of the instructors. Perhaps the instructor may think an exercise too long, too short, too difficult, too easy, or just not very instructive. If so, change it in any way you please. If it does not please you to do so, however, take comfort in the fact that most of the exercises have undergone an extensive evolution to arrive at the forms they now possess. It should be possible to get along fairly well with any of them.

Finally, those exercises that require the use of heavy, bulky, or immobile lab equipment, and which should be conducted exclusively in the most permanent lab room are marked with asterisks in the table of contents. Otherwise, anything goes.

I would like to thank Bill Schuster, John Pratt, Steve Gregory, Rich Allen, Alice Hine, Ross Shuart, Leo Connolly, Pete Mantarakis, Carolyn McCarthy, Bob Dukes, Ray White, and all others who in 1973 wrote, rewrote, or adapted the original versions of the lab exercises in this book for use in the University of Arizona Astronomy 1a and 1b labs; Bob Wallis and Dave Proctor at Baldwin-Wallace College for their support of Astronomy at Baldwin-Wallace; and the following reviewers for their helpful comments on how to make the labs in this book desirable and useful to the greatest number of colleges: Harry Augensen, Widener University; Ann Cowley, Arizona State University; Roy Garstang, University of Colorado at Boulder; Neil Lark, University of the Pacific; and Douglas Nagy, Ursinus College. I would like to thank Nick Sanduleak and Peter Pesch of Case Western Reserve University for the use of an objective prism plate and for helpful advice and encouragement. Many thanks to a terrific Wadsworth team for making this book a reality. I also wish to thank my wife, Mary Ferguson, for putting up with a husband who sometimes puts astronomy ahead of more important things. Much of value in this text is the work of others. Any errors are mine.

Dale C. Ferguson

Extra Equipment and Materials Needed for the Exercises

EXERCISE NUMBER	Equipment Needed and Notes
1	Models or examples of telescope mountings and accessories (helpful but not essential).
2	A few small telescopes, star charts.
3	Meter sticks, calculators with square roots keys, a star photograph to be measured, centimeter rulers.
4	Calculators with square root keys. A calculator or computer programmed with the vis-viva equation is helpful (but not essential) for this one.
5	Celestial globes, rulers.
6	Calculators with square-root keys.
7	Telescope(s) with setting circles, sidereal clock.
8	Celestial globes, sidereal clock, telescopes, *Atlas Eclipticalis* or *Atlas Borealis* (both by Antonin Becvar of the Prague Observatory), or *Sky Atlas 2000.0* (by W. Tirion, published by Sky Publishing Co., Cambridge, Mass., 1981), if available.
9	Ray-tracing device with lens and prism cross-section pieces, lenses of different focal lengths, optical benches, light bulbs (with patterns put on in dry mark), meter sticks, concave mirrors.
10	Long telescope (refractor preferred), ground glass, hot plate; stars should be visible.
11	Large wall map of moon, lunar globe if available, rulers.
12	Telescopes, large moon maps; moon *must* be visible.
13	No extra equipment.
14	*Observer's Handbook* (published annually by the Royal Astronomical Society of Canada) or *Astronomical Almanac* (published annually by the U.S. Naval Observatory); telescope(s), stop watch if available. At least one planet must be visible.
15	Telescope, sun filter or projection screen, hand-held spectroscopes if available. Best if done at midday or morning; sun must be visible.
16	Rulers, protractors, simple sextants, calculators, light table if available.

EXERCISE NUMBER	Equipment Needed and Notes
17	Calculators with square-root keys.
18	UBV photoelectric photometer, telescope, calculator with base 10 logarithms (or log tables); clear steady night.
19	Laboratory spectroscope, hand-held spectroscopes, transmission diffraction gratings, emission tubes and power supplies, spectroscope with salt burner if available, star spectroscope, telescope. May be done as an indoor lab if stars or star spectroscope not available.
20	No extra equipment.
21	Light tables if available, logs to base 10 optional.
22	Telescope with equatorial mounting, clock drive, and setting circles, star atlases if available, sidereal clock, *Observer's Handbook* or *Astronomical Almanac;* celestial objects must be visible.
23	Calculators.
24	Calculators, logs to base 10 optional.
25	Image processing software.
26	Clock-driven telescope with CCD camera. CCD camera software and computer to run it; *must* be clear night.
27	Magnifiers and/or *Hubble Atlas* (published by Hale Observatories), if available.
28	Calculators.
Appendix A	Binoculars and/or small telescopes, star charts.
Appendix B	Binoculars and/or small telescopes, star charts; dark sky *necessary,* best if done in October.
Appendix C	Binoculars and/or small telescopes, star charts.
Appendix D	Binoculars and/or small telescopes; dark sky *necessary,* best if done in March.

Name: _____

Exercise One: Indoor

Introduction to Astronomical Telescopes

I. Types of Optical Telescopes

The Refracting Telescope

This type of telescope uses a lens to collect and focus the light. The image is formed behind the objective lens at the opposite end of the telescope tube.

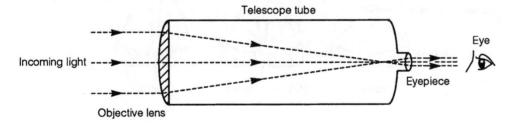

The Reflecting Telescope

This type uses a concave primary mirror to collect and focus the light. One or more secondary mirrors may be used to place the observation point in a convenient location. These secondary mirrors are usually flat or convex.

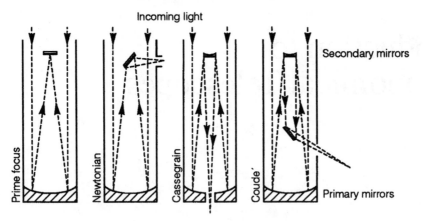

Systems That Use Both a Lens and a Mirror

In these telescopes a correcting lens is placed in front of the primary mirror. This arrangement allows the use of primary mirrors with more simply shaped reflecting surfaces. Also, other desirable properties, such as a larger undistorted field of view, can be obtained.

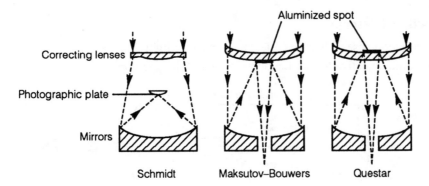

II. Types of Telescope Mountings

Altazimuth Mounting

This type of mounting is very easy to build and set up because one axis is vertical and the other is horizontal. However, you must rotate the telescope about both axes to follow a star. Also, you cannot observe near the zenith. One type of altazimuth mounting, called the Dobsonian, allows observation near the zenith, and is becoming increasingly popular with amateur astronomers.

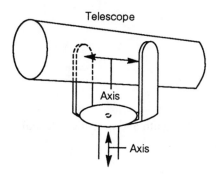

German Equatorial Mounting

This mounting is harder to set up because one of the axes must be parallel with the Earth's axis of rotation. However, you need to rotate only the one axis to follow a star.

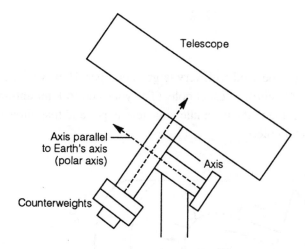

Exercise One: Introduction to Astronomical Telescopes

Fork Mounting

This is also an equatorial type mounting—you rotate only one axis to follow a star. This type of mounting makes it easier to observe near the zenith.

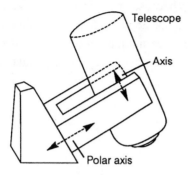

English Mounting

With this equatorial mounting it is easy to observe at the zenith and near the celestial pole.

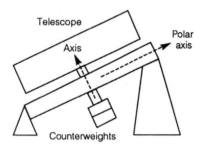

Yoke Mounting

This is an equatorial mounting that can be used with very large telescopes. However, with this mounting you cannot look near the north celestial pole. (The yoke and fork mountings may be combined to produce a mounting that can handle large telescopes and that allows you to observe near the north celestial pole.)

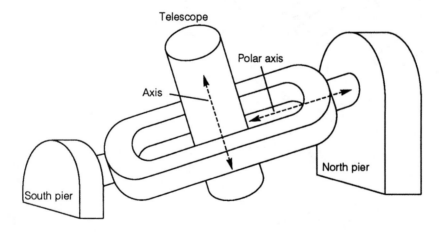

Exercise One: Introduction to Astronomical Telescopes

III. Telescope Accessories

- *Eyepieces:* Lens systems used to make visual observations. The eyepiece magnifies the image produced by the objective of a telescope. Usually a number of eyepieces are available, each producing a different magnification.

- *Camera:* This can be used to take photographs through the telescope.

- *Finder telescope:* A small telescope (with a wide field of view) that is mounted on the side of the main telescope. The finder telescope usually has cross hairs in its field of view so, if properly aligned, it can aid in finding astronomical objects with the main telescope.

- *Clock drive:* A motor that slowly rotates the telescope about one of the axes of the mounting. In this way the telescope automatically follows the object being observed.

- *Slow motion controls:* Controls allowing the observer to rotate the telescope manually about either of the axes of the mounting.

- *Setting circles:* Graduated metal circles attached to the two axes of the telescope mounting. When the telescope mounting is properly aligned, these circles make it much easier to locate objects in the sky.

- *Solar observing screen:* A white screen upon which an image of the sun can be projected. In this way the sun can be harmlessly viewed with a telescope.

- *Filters:* Very dense filters can be used with a telescope in order to observe the sun directly. Also, color filters can be used to bring out details on the planets.

- *Photometers, spectroscopes, polarimeters, filar micrometers, image tubes:* Some of the more sophisticated pieces of equipment that professional and serious amateur astronomers can use with a telescope.

IV. Terms Used to Describe the Performance of a Telescope

- *Light-gathering power:* The ability of a telescope to collect more light per unit time than the pupil of the eye does. The light-gathering power depends upon the diameter of the objective of the telescope.

- *Resolving power:* The capability of a telescope to separate or resolve two stars that are so close together in the sky that the unaided eye sees them as a single point. The resolving power is also determined by the diameter of the telescope's objective.

- *Magnification or magnifying power:* The number of times larger (in angular diameter) an object appears through a telescope than to the naked eye.

- *Scale:* The linear distance in the focal plane of a telescope that corresponds to a particular angular distance in the sky—for example, so many centimeters per degree.

Exercise One: Introduction to Astronomical Telescopes

- *Focal length:* The distance between a lens or mirror and its focus.
- *Focal ratio or F-ratio (speed):* The ratio of the focal length of a telescope's objective to its diameter. For an extended object such as the moon, a planet, a nebula, or a galaxy, the f-ratio determines the brightness of its image.

Name: _____

Exercise Two: Outdoor

Constellations and the Celestial Sphere

I. Definitions

- *Celestial sphere:* An imaginary sphere of very large radius that surrounds the observer. For many astronomical purposes, celestial objects are considered as if they were all located on the surface of this sphere.

- *Zenith:* The point in the sky that is directly above the observer.

- *Horizon:* The great circle on the celestial sphere that is 90° from the observer's zenith in every direction.

- *Altitude:* The angular distance of a celestial body above the horizon, measured along a vertical circle. If an object is at the zenith, its altitude is 90°; at the horizon its altitude is 0°.

- *Azimuth:* The angular bearing of an object in the sky, measured from north (0°) through east (90°), south (180°), and west (270°) back to north (360° or 0°).

- *Latitude:* The north–south coordinate on the surface of the earth; the angular distance north or south of the equator measured along a meridian passing through a place.

- *Longitude:* The east–west coordinate on the Earth's surface; the angular distance, measured east or west along the equator from the Greenwich meridian to the meridian passing through a place.

- *Celestial poles:* The two points at which an extension of the Earth's axis intersects the celestial sphere.

- *Celestial equator:* The projection of the Earth's equator onto the celestial sphere. The celestial equator is 90° from the celestial poles in every direction.

- *Magnitude:* A measure of the brightness of a celestial body. The brighter the object the lower the magnitude; thus the bright star Aldebaran is of magnitude 1, and the faintest stars normally visible to the naked eye are of magnitude 6.

II. Motions of the Stars

1. At the beginning of the lab, note the position of the following: (a) Polaris (see Part III), (b) a star just above the eastern horizon, (c) a star just above the western horizon, and (d) a star about 45° above the southern horizon. Pick bright, conspicuous stars and try to indicate the positions in terms of altitude and azimuth. When measuring angles on the celestial sphere, keep in mind that one-fourth of a circle is 90° and that the distance between the "Pointers" of the Big Dipper is about 5°.

2. Near the end of the lab session, note the new positions of the same stars to find out what motion has taken place.

3. What can you conclude?

III. Polaris and the Observer's Latitude

1. Find Polaris. This can be done most easily using the "Pointers" of the Big Dipper. As their name suggests, a line drawn through these two stars points at Polaris.

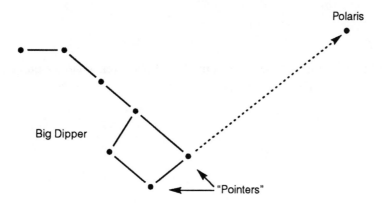

If the Big Dipper is not easily visible, you can use the constellation Cassiopeia to locate Polaris. Cassiopeia is a fairly conspicuous constellation shaped like a W (or M). If one views Cassiopeia as an upright W, then the Little Dipper is directly above the W. Polaris is the star on the end of the handle of the Little Dipper. Under unfavorable circumstances (such as with city lights or a hazy sky), you may be able to see only Polaris and the stars β and γ in the Little Dipper.

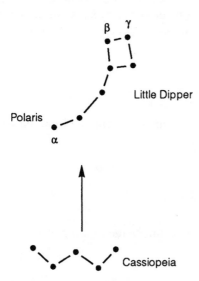

Exercise Two: Constellations and the Celestial Sphere

2. Find Polaris with an equatorially mounted telescope. (Remember, an equatorial mounting has one axis parallel to the Earth's axis.) Now note the position of the telescope with respect to the axis of the mounting. What does this tell you about the position of Polaris? (Hint: See the definition of celestial poles.)

3. Next estimate the altitude of Polaris. How is the altitude of Polaris related to the observer's latitude? Draw a diagram showing this relationship.

IV. Limiting Magnitude

1. Look around the sky and find a star that is as faint as the faintest stars you see.

2. Draw a map showing this star and its location with respect to nearby stars. Show enough bright stars that you can identify the constellation from the star map.

3. Find your faintest star on a detailed star map and determine its magnitude from the size of the dot on the map and the magnitude key given on the map.

4. Alternate method: Use the stars of the Little Dipper to estimate the limiting magnitude.

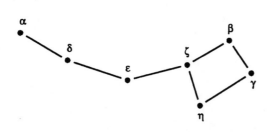

Star	Magnitude
α	2
β	$2\frac{1}{4}$
γ	$3\frac{1}{4}$
δ	$4\frac{1}{2}$
ε	$4\frac{1}{2}$
ζ	$4\frac{1}{4}$
η	5

5. What factors affect this limiting magnitude? Why is your answer different from the value of sixth magnitude, which is normally quoted?

V. Constellation Study

1. Throughout the evening, your lab instructor will be around to point out the constellations visible this time of year.

2. Using star charts, try to pick out as many of the constellations as you can on your own.

Exercise Two: Constellations and the Celestial Sphere

Name: _____

Exercise Three: Indoor

Introduction to Experimental Measurements

I. Significant Figures

Significant figures are defined as those figures in a number that contain meaningful information in view of the error or uncertainty involved in obtaining the numerical result. In general, all figures listed as part of a decimal number should be significant except for zeros that must be added to define the location of the decimal point. Such zeros are therefore present only because of the system of units being used in the expression and do not indicate any additional precision in the measurement. The significant figures in the following numbers are underlined:

a. .00<u>27</u>   b. <u>1</u>.<u>4003</u>   c. <u>100.0</u>

d. <u>1</u>.<u>43</u>   e. <u>42</u>,000

1. Underline the significant figures in the following numbers:

a. 1.0003   b. 2.70   c. .042

d. 1007.030   e. 2,700

The result of an addition or subtraction should be rounded off so that the final answer has significant digits only in place locations that are equal to or larger than the lowest place

Exercise Three: Introduction to Experimental Measurements

value containing a significant figure in all of the numbers used to form the sum or difference. The following examples will illustrate the technique:

a. 4.002 + 3.2 = 7.2~~02~~

b. .006 + 5.10 = 5.1~~06~~ = 5.11

c. 120.0 + 10.21 = 130.2~~1~~

d. 11.05 + .037 = 11.0~~87~~ = 11.09

2. Perform the following calculations, paying attention to significant figures in the final answers:

a. 1400.3 + 70.1 = _____

b. 6007 + 2.74 = _____

c. .0061 + .00037 = _____

In multiplication and division, the final answer should always be given to the same number of significant figures as the *least* accurate figure used in the calculation. The following product will serve as an example:

$$4.02 \times 2.1 = 8.4~~42~~$$

3. As a demonstration of the errors that may result from not observing significant figures in such an answer, calculate the following products without regard to significant figures:

a. $4.024 \times 2.14 =$ _____

b. $4.020 \times 2.10 =$ _____

c. $4.015 \times 2.05 =$ _____

4. On the basis of these calculations, explain how the concept of significant figures could be of importance in scientific research.

II. Scientific Notation

Most numerical results can best be expressed in scientific notation. A numerical result written in this notation takes the form of a decimal number from zero to ten that is multiplied by a factor consisting of the number ten raised to an integer power. As a rule only the significant figures of the original number are retained as part of the decimal. Thus zeros that serve only to indicate the position of the decimal point are dropped in scientific

Exercise Three: Introduction to Experimental Measurements

notation, and the decimal position is determined by the power of ten. Consider the following examples:

a. $0.004 = 4 \times 10^{-3}$ (4 times 10 to the minus 3rd)

b. $24 = 2.4 \times 10^{1}$ (2.4 times 10 to the 1st)

c. $1,600,000 = 1.6 \times 10^{6}$ (1.6 times 10 to the 6th)

1. Measure the length of the lab table and record the length in centimeters, using scientific notation.

2. Estimate the size of the error that may be present in your measurement. How large is this error in percent of the overall length measured?

3. How would you write the length of the table in meters instead of centimeters?

III. The Calculator

The instructor will demonstrate the use of the calculator for doing simple calculations. The remainder of the lab should then be completed using a calculator as much as possible.

1. Perform the following operations with a calculator. Rewrite your answers giving attention to significant digits.

a. $2.4 \times 10.6 =$ _____   Rewritten: _____

b. $3.48/12 =$ _____   Rewritten: _____

c. $\sqrt{34} =$ _____   Rewritten: _____

d. $(10.2)^{2} =$ _____   Rewritten: _____

Exercise Three: Introduction to Experimental Measurements

IV. The Normal Curve of Error

The errors associated with the results of a particular experiment can be reduced by repeating the same experiment many times over. The average value of the results is then much more meaningful than the result of any one of the individual experiments. The treatment of data resulting from such experiments can best be evaluated by using the techniques of statistical analysis that predict that the numerical results will be spread out evenly around the average or mean value, which can be denoted by $\bar{x}$. This mean value is equal to the sum of the individual experimental values of x divided by the number of experiments used to form the sum. The following example shows the results of a typical experiment that was repeated four times:

Time 1: $x = 4.2$ Time 3: $x = 4.4$

Time 2: $x = 3.8$ Time 4: $x = 5.6$

1. Determine the average of the results from the four times that the experiment was repeated.

Performing this experiment many more times will eventually lead to a distribution of the experimental data in the form of a Gaussian or normal curve of error. Such a curve is created by plotting the number of repetitions that result in a particular experimental value of x as a function of x itself. The general shape of such a normal curve of error is always the same, regardless of the nature of the experiment, as long as there are no physical conditions present that might place restrictions on the range of x near its mean value. A normal curve of error is shown in Figure 3.1. The width of the curve is characterized by the standard deviation, which is a measure of the spread of the experimental values about the mean. Specifically, the standard deviation is found by calculating the individual differences between the experimental values and the mean $\bar{x}$. Each of these differences is then squared, and they are averaged. The square root of the average taken in this way is the standard deviation. The standard deviation is a useful quantity because 68% of the experimental results should be within one standard deviation away from the mean value $\bar{x}$ if the data are ideally distributed.

Exercise Three: Introduction to Experimental Measurements

Figure 3.1
Normal curve of error.

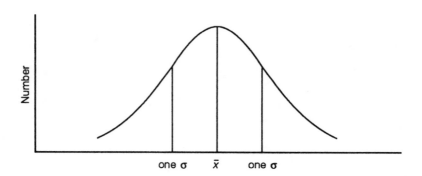

2. Calculate the standard deviation of the four experimental results that were given on page 16.

3. The centimeter rulers will now be employed to determine experimentally the position of a star on a photograph supplied by your instructor. The instructor will select a suitable star for measurement and explain the procedure. Estimate the position to within 0.1 mm. Make a histogram of the data obtained by the class by plotting the number of measurements as a function of the star's position on the photograph. Estimate the standard deviation of the data by determining the spread in position about the mean that would be just large enough to include 68% of the measurements.

Name: _____

Exercise Four: Indoor

Interplanetary Travel

In this lab you will take part in an imaginary manned flight to Mars. In so doing, you will learn about orbits, periods, and velocities of natural planets and spacecraft.

I. Planetary Orbits and Periods

Below is a sketch of a planet's orbit around the sun. Kepler's first law tells us that it is an ellipse, with one focus at the sun: a, the semimajor axis, is the average of the perihelion (closest to the sun) and aphelion (farthest from the sun) distances of the planet. The period of a planet (its year) is given by the formula $P = \sqrt{a^3}$, where the period P is given in years and a is given in astronomical units (AU).

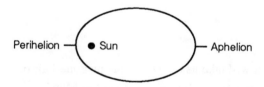

1. Given the perihelion and aphelion distances of the following planets, calculate the semimajor axis of each planet's orbit (in AU) and the length of its year, or period (in Earth years).

	Perihelion distance (AU)	Aphelion distance (AU)	a (in AU)	P (in years)
Mercury	.31	.47		
Venus	.720	.726		
Earth	.98	1.02	1	1
Mars	1.38	1.66		
Jupiter	4.95	5.46		

II. Transfer Orbits

The orbit of the manned spacecraft we will send to Mars will also be an ellipse, with the Earth's orbit as perihelion and Mars' orbit as aphelion. This configuration takes the least amount of energy (requires the least fuel).

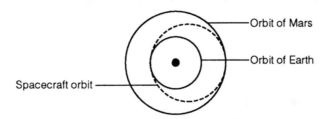

1. We will try to launch our spacecraft so that when we reach Mars it will be at perihelion. What would be the perihelion distance, aphelion distance, a, and P of our spacecraft? (Assume the Earth's orbit is circular.)

2. Since we want to stop at Mars, the time it will take us to get there is just one-half our orbital period around the sun. How long (in days) will it take us to get to Mars?

Figure 4.1
Earth and Mars on launch and arrival dates.

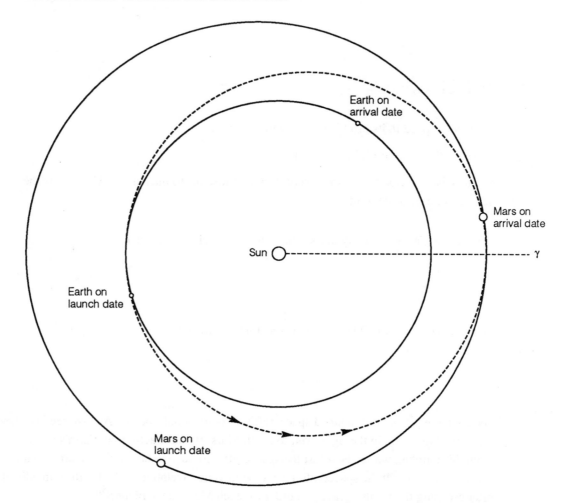

3. We decide to make our trip in 2016 A.D. To reach Mars on October 27 (the approximate date it reaches perihelion that year), on what date would we have to leave Earth?

4. We find that we can't reach Mars on October 27, 2016, because on our calculated launch date the Earth is not diametrically opposite the sun from where Mars will be October 27 (our condition for minimum energy). Instead we launch on the first day of April. Now when we reach Mars it will be 1.41 AU from the sun. Because Mars will not be at perihelion when we arrive, how much longer will the trip take than we had calculated before? (Hint: Spacecraft aphelion distance has changed.)

Exercise Four: Interplanetary Travel

5. What will be the historic date of our landing?

III. Velocities

A planet's speed in its orbit is given by the formula

$$V^2 = (30 \text{ km/sec})^2 \cdot [(2/r) - (1/a)]$$

where V is the velocity, r is the distance (at that instant) to the sun in AU, and a is the semimajor axis of its orbit.

1. What is the Earth's velocity in its orbit? (Assume a circular orbit.)

2. What is the velocity of Mars when it is at 1.41 AU from the sun?

3. By calculating what our manned spacecraft's velocity should be at 1 AU, we see how fast it must be going to be in the right orbit to reach Mars. By subtracting the Earth's velocity from this number, we see how fast the spacecraft will have to be moving relative to the Earth. How fast will the spacecraft have to be traveling (relative to the Earth) immediately after escaping the Earth's gravity in order to reach Mars as we planned?

$$V_{rel} = V_{spacecraft} - V_{planet}$$

4. How fast will the spacecraft be moving relative to Mars when we arrive there at 1.41 AU?

When the spacecraft reaches Mars, will it need to speed up or slow down to match Mars' velocity?

Exercise Four: Interplanetary Travel

IV. Communicating with Earth

1. While we are on Mars, we will want to radio Earth now and then. At our farthest point from Earth, Mars will be about 2.66 AU from Earth. Knowing that $1 \text{ AU} = 1.5 \times 10^{13}$ cm and that our radio signals travel with the speed of light, 3×10^{10} cm/sec, how long will it take our radio messages to reach Earth at its farthest point?

2. If we were trying to carry on a conversation with Earth, how long would we have to wait to hear the answer to an easy question we asked someone on Earth?

V. Return

We can't just return anytime we'd like to. We'll have to wait to take off (in order to conserve fuel) until the Earth will be in the proper position at the end of our transfer orbit home. In general, we will have a long wait, but we'll carry enough provisions to make it through (and we can explore Mars in the meantime).

1. An accurate calculation of the time we will have to wait is complicated, but we can put an upper limit on it by realizing that after one synodic period the Earth will be in the same place, relative to Mars, as it was when we landed. Surely the two planets will go through the proper configuration for launch before then. Using the formula

    $$1/P_{syn} = (1/P_{Earth}) - (1/P_{Mars})$$

    calculate what the synodic period (and thus the longest wait time) would be. (P_{syn} is the synodic period.)

2. We have finally arrived home with samples, photos, and so on. Assuming we had to wait 3/4 synodic period before leaving Mars, how long in all did the trip take?

Exercise Four: Interplanetary Travel

IV. Communication with Earth

V. Actions

Name: _____

Exercise Five: Indoor

The Seasons

I. Finding the Relative Amounts of Sunshine You Receive at the Summer and Winter Solstices

1. Set the celestial globe for your particular latitude. That is, the north celestial pole should be your latitude angle above the northern horizon. Locate the sun's position at the summer solstice (its declination will be 23 1/2°).

2. Measure the altitude of the sun according to the celestial globe at half-hour intervals from sunrise to sunset. Take advantage of symmetry about the meridian to save work. Also, record the azimuth and hour angle (see Exercise Eight) of the setting sun.

3. Find out how much land area a beam of sunlight is spread over when the sun is at each altitude compared to the area when the sun is directly overhead. See the diagram that follows.

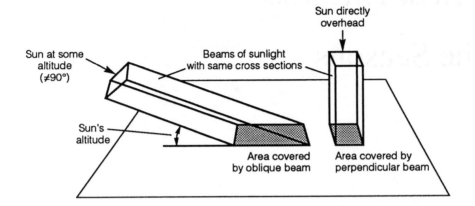

To calculate the relative areas covered, you can construct a diagram as shown in the next diagram and measure X and L with a ruler for altitudes (A) of 0°, 10°, 20°, 30°, 40°, 50°, 60°, 70°, 80°, and 90°. (You can also use trigonometry.)

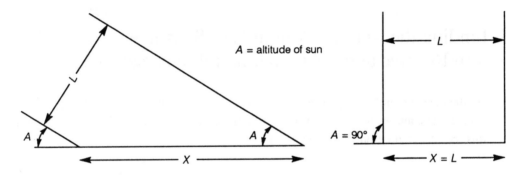

The ratio X/L gives the amount by which the sun's rays have been spread out because the sun is not directly overhead. Thus the ratio L/X gives the reduction of the amount of sunlight for a given area of land compared to the amount of sunlight if the sun were overhead. That is, because the sunlight per unit land area is inversely proportional to the area covered by a unit beam,

$$\frac{\text{sunlight with sun at some altitude}}{\text{sunlight with sun at zenith}} = \frac{L}{X}$$

4. Plot a graph of the ratio *L/X*, using the values measured in Question 3, using the grid at the end of the chapter. Draw a smooth curve through the points. For example, see the diagram that follows.

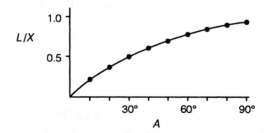

L/X = Relative sunlight

A = Altitude of sun

5. Use the curve you plotted in Question 4 to obtain the relative sunlight for each of the altitudes you recorded in Question 2.

6. Add up these relative sunlight values to get the total sunlight received over the entire day. (You may wish to correct for the fact that the last value before sunset may not represent a full half-hour of sunshine.)

7. Repeat all of the above steps with the sun at the position of winter solstice (declination of –23 1/2°). (Obviously, for steps 3 and 4 you can use the graph you derived earlier.)

8. Find the ratio of the amount of sunshine at summer solstice to the amount at winter solstice.

Exercise Five: The Seasons

II. Questions You Can Answer with the Help of the Celestial Globe

1. How many hours of "sun-up" do you have at the solstices and equinoxes? How does this compare with Tucson, Arizona (32° N)?

2. How does the sunshine ratio (from Question I.8) compare with the "sun-up" hours ratio for your location? From this information, tell which gives us our biggest seasonal effect, the obliquity of the sunlight or the number of hours of sunshine.

3. Compare the azimuths of the setting sun at the solstices and equinoxes for Tucson and for your location.

4. Twilight is defined astronomically to end when the sun is 18° below the horizon. Does it last longer at the winter or summer solstice where you are located? Where does it never end at the summer solstice?

5. When does a place on the equator receive the most and the least sunshine? Describe the seasons at the equator and in the southern hemisphere.

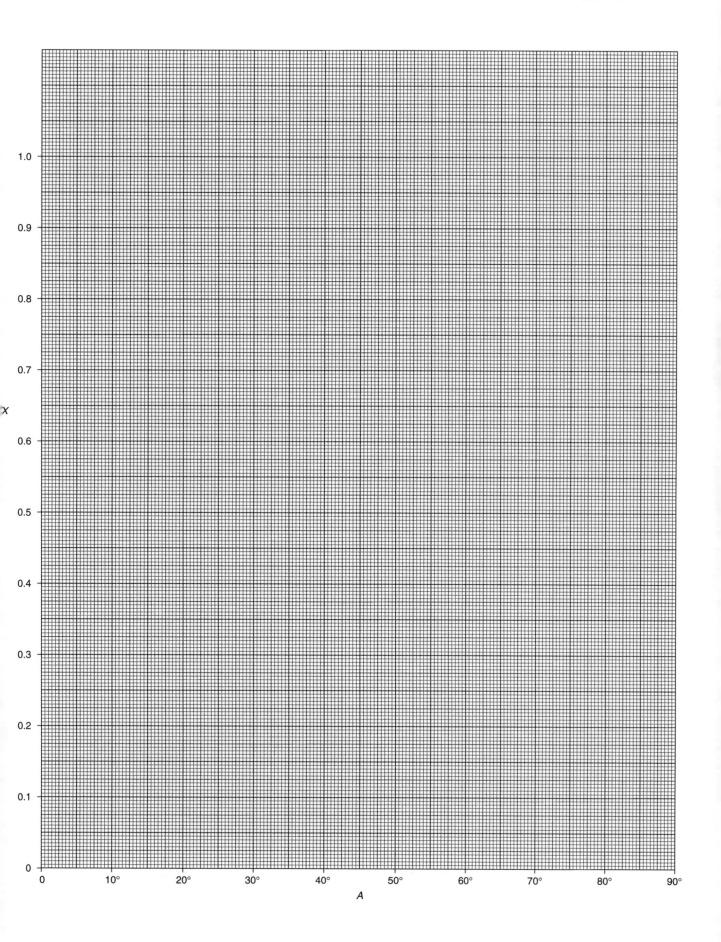

Name: _____

Exercise Six: Indoor

The Temperature of the Earth

I. Introduction

Planets gain heat by absorbing light from the sun and lose heat by radiating it away into space as infrared radiation. A planet's temperature depends on the amount of heat energy it absorbs and radiates, as well as on internal heat sources (such as radioactivity and contraction). The atmosphere of a planet that has one can significantly increase the surface temperature, producing conditions quite different from those the planet would have without an atmosphere. In this lab, you will use simple assumptions and well-known quantities to determine what the average surface temperature of the Earth would be if it had no atmosphere, and you will learn to appreciate the "greenhouse effect."

II. Solar Luminosity

Stars like the sun radiate, to a good approximation, like black bodies (hypothetical bodies that absorb and emit radiation perfectly). For instance, the sun radiates the same amount of energy as a spherical black body, of the same size as the sun, having a temperature of 5800 K. (The Kelvin temperature scale uses degrees that are the same size as those in the Celsius scale, but it starts at absolute zero, –273°C.) The sun's radius (R) is 6.96×10^{10} cm.

1. Calculate the surface area of the sun, in square centimeters, using the following formula:

   $A = 4\pi R^2$

   Place your answer here:

   $A =$ _____ cm^2

   The Stefan–Boltzmann law for the luminosity (L), the energy radiated by a black body of temperature T, is:

   $L = \sigma A T^4$

   Here, the Stefan–Boltzmann constant $\sigma = 5.67 \times 10^{-5}\,erg/cm^2\text{-}K^4\text{-sec}$.

2. Using this law, calculate the luminosity of the sun in ergs per second.

   $L =$ _____ erg/sec

III. The Solar Energy Absorbed by the Earth

Although the sun radiates a very large amount of energy per second, as you just calculated, only a small amount crosses each square centimeter of area at the distance of the Earth from the sun. Assuming that the entire sun's luminosity crosses a sphere with a radius as big as the Earth's orbit around the sun you can calculate how much will hit each square centimeter of surface of this imaginary sphere each second (the solar flux, F).

1. Take the distance of the Earth from the sun (d), which is 1 AU, to be 1.5×10^{13} cm, and calculate F, using the following equation:

   $F = L/(4\pi d^2)$

   $F =$ _____ $erg/cm^2\text{-sec}$

   Of the area of the large imaginary sphere, only the cross-sectional area of the Earth will intercept the light from the sun. That is, of the sun's luminosity, the Earth will intercept

Exercise Six: The Temperature of the Earth

only an amount given by multiplying its cross-sectional area by the flux you just found. Thus,

$$E = F \times \pi r^2$$

where r is the radius of the Earth, 6.38×10^8 cm.

2. Using this formula, calculate the amount of solar energy E hitting the Earth each second.

$E =$ _____ erg/sec

Of course, not all of this energy is absorbed by the Earth. Some of it is reflected back into space. The fraction of the solar energy reflected is called the Earth's albedo, a, which has been measured (by observing the brightness of the earthlit portion of the new moon) to be about 0.35. Thus, the fraction absorbed (the absorptance) is $\alpha = (1 - a) = 0.65$.

3. Multiply the E you just obtained by this α to obtain the solar energy absorbed by the Earth each second, L_e, and write the answer here:

$L_e =$ _____ erg/sec

IV. The Earth as a Black Body

Convection and conduction can't cool planets, because of the emptiness of space. Most solid bodies emit infrared radiation in about the same amount as a black body of the same size and temperature. In the absence of any internal sources of heat, the temperature of a solid planet like the Earth will become such that it will radiate away as infrared radiation the same amount of energy it absorbs from the sun. If we can assume that it radiates like a black body, we can calculate the temperature at which that will happen, and this will be the temperature of the planet. In our calculations, because we are interested only in an average temperature, we will assume that the solar energy the Earth absorbs is quickly distributed around the surface by the Earth's rapid rotation (a questionable assumption). Then, we can take the energy absorbed from the sun, L_e, use it as the energy radiated by the whole Earth, and turn the Stefan–Boltzmann equation around to derive an average temperature, T_e, for the Earth:

$$T_e^4 = L_e/(4\pi r^2 \sigma)$$

Exercise Six: The Temperature of the Earth

1. Using this formula, find the fourth power of the Earth's temperature, and write the answer here:

    $T_e^4 = $ _____ K^4

2. Now take the square root of this twice to find the average temperature of the Earth:

    $T_e = $ _____ K

3. Subtract 273 from your answer to change it to degrees Celsius:

    $T_e = $ _____ °C

4. Water freezes at 0°C and boils at 100°C. According to your just-derived T_e, could liquid water exist on the Earth's surface? Why or why not?

5. Considering that the Earth's surface is mostly covered with liquid water, how do you think your derived temperature compares with the true average temperature on the Earth?

V. The Atmosphere's Influence: The Greenhouse Effect

The primary reason your derived temperature in the previous section is in error is the influence of the atmosphere on the Earth's temperature. Water vapor and carbon dioxide (CO_2) in the atmosphere keep the infrared radiation emitted by the surface from escaping into space and thus hold the heat in. This is the reason that, all other things being equal, a cloudy night will be warmer than a clear night. The water vapor in the clouds keeps the infrared radiation from being lost into space. This is called the greenhouse effect. In a greenhouse, as for a planet, convection and conduction are minimized. The glass in the

Exercise Six: The Temperature of the Earth

greenhouse lets light in, but it keeps the infrared radiation emitted by the surfaces inside from escaping, thus warming the greenhouse. Although clouds in the Earth's atmosphere can make the nights warm, they can also make the days cool by blocking the visible light from the sun. However, the CO_2 in the atmosphere is transparent to visible light but opaque to infrared, and so for the Earth CO_2 takes the place of the greenhouse glass.

The fraction of the infrared energy emitted by a body compared to that emitted by a black body of the same temperature is called the emittance, epsilon. We can find the Earth's average emittance by using the Earth's real temperature, T_r, and modifying the Stefan–Boltzmann equation to find the Earth's emittance:

$$\varepsilon = L_e/(4\pi r^2 \sigma T_r^4)$$

1. Calculate ε for the Earth from this formula, assuming that the real Earth temperature $T_r = 12°C = 285$ K. Write your answer here:

$\varepsilon = $ _____

The value of ε for the Earth depends to a great extent on the amount of CO_2 in the atmosphere. If an increase in the CO_2 in the atmosphere decreased ε by 1.5%, it should make a difference of about 1°C in the Earth's average temperature. This could lead to increased melting of the polar caps, higher sea levels, warmer summers and winters, and a long-lasting change in the Earth's climate.

2. How can we help prevent the amount of CO_2 in the atmosphere from increasing?

Exercise Six: The Temperature of the Earth

VI. Venus and Mars: Contrasting Extremes of the Greenhouse Effect

For a planet that is slowly rotating or for which the greenhouse effect is not important, and that has an emittance equal to its absorptance, we can calculate a theoretical subsolar (high noon) temperature (T_{ss}). Equating the luminosity of the solar black body to that of a black-body sphere with a radius as big as the planet's orbit and performing a little algebra,

$$T_{ss}^2 = (R_{sun}/d_{planet}) \times T_{sun}^2$$

Because Venus rotates slowly, this is the temperature that the surface directly underneath the sun should reach, in the absence of an atmosphere.

1. Taking $d_{Venus} = 1.08 \times 10^{13}$ cm, $R_{sun} = 6.96 \times 10^{10}$ cm, and $T_{sun} = 5800$ K as we had before, calculate T_{ss} for Venus:

   T_{ss} (Venus) = _____ K

2. How does this compare with the 700 K temperatures measured by Venus probes on the surface of Venus?

3. Venus has a thick atmosphere made up mainly of CO_2. How can you account for the high surface temperature of Venus?

4. Mars has a thin atmosphere that is, however, mostly CO_2. Use the formula at the beginning of the section, with an average distance of Mars from the sun of 2.28×10^{13} cm, to calculate a T_{ss} for Mars, and place your answer here:

   T_{ss} (Mars) = _____ K

Exercise Six: The Temperature of the Earth

5. How does your calculated temperature compare with the observed subsolar temperature of Martian soil of 300 K?

6. Is the greenhouse effect apparently important on Mars?

Although the subsolar surface temperature on Mars would be quite pleasant, the average temperature for the planet's surface is about 212 K, far below the 273 K freezing point of water. However, photos have clearly shown that there are dry riverbeds on Mars, so water must have been liquid there at some time in the past.

7. Given the fact that Mars has extensive polar caps of frozen water and frozen CO_2, can you guess how liquid water could have existed on Mars at times in the past?

Exercise Six: The Temperature of the Earth

Name: _____

Exercise Seven: Outdoor

Astronomical Systems of Time

This exercise shows the relationships between the fixed and rotating equatorial coordinate systems. You will observe the sidereal time and compute the local solar and standard zone times. Before you go outside to make the necessary observations, your instructor will explain the rotating coordinate system and the motion of the sun on the celestial sphere. Figure 7.1 illustrates some of the relations involved.

Figure 7.1
Rotating coordinate system.

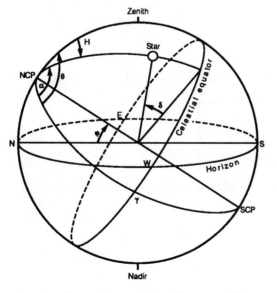

NCP = North celestial pole

SCP = South celestial pole

φ = Observer's latitude
= Altitude of NCP

α = Right ascension (RA) of star

θ = Right ascension of meridian
= Sidereal time

H = Hour angle of star

δ = Declination of star

γ = Vernal equinox (first point of Aries)

Exercise Seven: Astronomical Systems of Time

I. Sidereal Time (ST)

1. Using a telescope, find the hour angle of three of the stars listed. Record your observations in the table. Be sure to note whether the hour angle is east (negative) or west (positive) and the standard time of your observations.

Star	RA (2010.0) α	Declination (2010.0) δ	Standard time	Hour angle (HA)	Sidereal time (ST)
Caph: β Cassiopeiae	00^{h}09^{m}44^s	+59°12'			
Mirfak: δ Persei	03^{h}43^{m}37^s	+47°49'			
Sirius: α Canis Majoris	06^{h}45^{m}33^s	−16°44'			
Regulus: α Leonis	10^{h}08^{m}56^s	+11°55'			
Spica: α Virginis	13^{h}25^{m}44^s	−11°13'			
Antares: α Scorpii	16^{h}30^{m}01^s	−26°27'			
Altair: α Aquilae	19^{h}51^{m}17^s	+08°54'			

2. Find the sidereal time from the equation

    ST = RA + HA

    and record it in the table.

3. Compute the average value of the standard time and ST in the table and record them here:

Standard time	
ST	

4. Note that to a reasonable approximation the sidereal time is just the hour angle of β Cassiopeia. Why is this so?

Exercise Seven: Astronomical Systems of Time

II. Local Solar Time

1. Your instructor will supply the RA of the sun for today's date from the Astronomical Almanac (published annually by the U.S. Naval Observatory). From this and the ST you just found, find the hour angle of the sun. This is defined as the local solar time. Record your answer here:

Right ascension of sun	
Local solar time	

III. Standard or Zone Time

Two corrections must be added to the local time to get standard time. These are the equation of time and the longitude correction.

The equation of time (Eq T) is defined as the difference between the right ascension of the mean sun (MS) and right ascension of the true sun (TrS). This varies from day to day, and your instructor will provide it.

$$\text{Eq T} = RA_{MS} - RA_{TrS}$$
$$= HA_{TrS} - HA_{MS}$$
$$\text{Eq T} = \underline{\hspace{2in}}$$

1. What does the sign attached to the value for the equation of time mean?

The longitude correction adjusts for the difference between the point of observation and the center of the time zone. For any location this correction is

Longitude Correction = (Longitude – Longitude of Time Zone Center)/(15°/hr)

Your instructor will calculate this quantity for you in hours, minutes, and seconds of time.

Exercise Seven: Astronomical Systems of Time

2. Enter the longitude correction, equation of time, and the hour angle of the sun (local solar time) here:

Standard time	
Equation of time	
HA of sun	

3. Now calculate the standard time from the equation

$$\text{Standard Time} = \text{HA of sun} + \text{Longitude Correction} - \text{Eq T} + 12^h$$

(Watch the signs!) Enter the answer here:

Standard time	

4. Compare the computed time with the standard time you recorded in Part I. How big is the error?

5. Discuss the sources of error.

IV. Discussion

1. The Andromeda Galaxy (M31) has a right ascension of $0^h\ 42.7^m$.

   a. What will be the ST when it crosses the meridian?

   b. What will be the hour angle of the sun at that time?

   c. What will be the standard time at that time?

Exercise Seven: Astronomical Systems of Time

d. What is the standard time now?

e. How far from the meridian is M31 now? East or west?

f. At what time will M31 reach the meridian tomorrow night? Explain.

g. The instructor will have one or more telescopes set on this object. Observe it with both the naked eye and through a telescope. Sketch position and appearance relative to nearby stars.

h. Record the time of your observation and the name of the instrument you used.

2. Define:

a. Right ascension.

b. First point of Aries (vernal equinox).

c. Mean solar time.

d. Sidereal time.

e. Longitude correction.

Exercise Seven: Astronomical Systems of Time

Name: _____

Exercise Eight: Outdoor

Observing with the Telescope, Part I: Locating Celestial Objects

I. Star Charts and Setting Circles

Star charts are the astronomer's maps of the sky. They show the stars and other objects whose positions on the celestial sphere change very slowly over the years. Objects in the solar system that move rapidly with respect to the distant stars can then be found by plotting their motions through the background of stars on the chart. Like maps of the Earth's surface, star charts are prepared according to different scales to show varying amounts of detail. Naked-eye charts are usually limited to stars brighter than sixth magnitude. Such stars can be seen with the unaided eye or with binoculars. The user of a small telescope, however, will need more detailed charts, such as those in Wil Tirion's *Sky Atlas 2000.0* (© 1981 Sky Publishing Corp.). This atlas has a limiting magnitude of 8.0, and the positions of even fainter star clusters, nebulae, and galaxies are included on the charts along with the stars. The *Atlas Eclipticalis* and *Atlas Borealis* by Antonin Becvar of Prague Observatory are fine examples of large-scale charts, and they contain virtually all stars that are easily visible with the aid of a small telescope. These charts also use a color code to indicate the spectra types and temperatures of the stars plotted. Look at all three of these atlases carefully, if available.

II. Star Names and Designations

The brightest stars on most charts are designated by Greek letters combined with the Latin genitive of the name of the constellation in which the star is located. Normally, the brightest star is given the letter α, the next brightest star β, and so on down through the alphabet. A notable exception is in Orion, where α was later found to be the second brightest star, and β is actually the brightest! Modern charts retain the Greek letter designations as well as a few of the more common names used by ancient sky observers. For example, the star Sirius is the brightest star in the constellation of Canis Major. It is also designated as α Canis Majoris.

Letters of the Greek Alphabet

α	Alpha	ι	Iota	ρ	Rho
β	Beta	κ	Kappa	σ	Sigma
γ	Gamma	λ	Lambda	τ	Tau
δ	Delta	μ	Mu	υ	Upsilon
ε	Epsilon	ν	Nu	φ	Phi
ζ	Zeta	ξ	Xi	χ	Chi
η	Eta	o	Omicron	ψ	Psi
θ	Theta	π	Pi	ω	Omega

III. Sky Coordinates

Most star charts have a gridwork of vertical and horizontal lines that indicate right ascension and declination. The lines of right ascension are drawn between the celestial poles, and the lines of declination are drawn around the celestial sphere parallel to the celestial equator. Just as you can pinpoint the location of any city on the Earth by stating its longitude and latitude, you can use the equatorial coordinates of a celestial object to locate it in the sky. Declination is the coordinate that gives the angular distance of an object north or south of the celestial equator. An object north of the celestial equator is indicated by a positive number of degrees, minutes, and seconds of arc, and the declination of an object south of the celestial equator is indicated by a negative number of degrees, minutes, and seconds of arc. The meridian position of an object on the celestial sphere is given by its right ascension. Right ascension is zero at the vernal equinox, which is the point in the sky where the sun crosses the celestial equator on the first day of spring. Right ascension increases in an eastward direction to a maximum of 24 hours, which represents one complete circuit of the sky. Displacing the telescope one hour in right ascension therefore corresponds to the movement of the stars that occurs in one hour of time because of the Earth's rotation.

Exercise Eight: Observing with the Telescope, Part I

IV. Sidereal Time

The last ingredient in locating sky objects is the local sidereal time. To find objects in the sky, it is easiest to remember that the sidereal time is always equal to the right ascension of the stars that are lined up exactly with the observer's meridian. The sidereal time therefore tells you directly what right ascension is lined up with the great circle that runs straight overhead from the north celestial pole to the south point on the horizon. The right ascension at the zenith is therefore also equal to the sidereal time at the observer's location. Figure 8.1 demonstrates how the lines of declination and right ascension might appear to an observer looking toward the south when the sidereal time is exactly $22^h\ 0^m\ 0^s$ in the fall or $10^h\ 0^m\ 0^s$ in the spring. One hour later the sky would appear as is shown in Figure 8.2. The sidereal time would then be exactly $23^h\ 0^m\ 0^s$ in the fall or $11^h\ 0^m\ 0^s$ in the spring.

V. Locating a Celestial Object

1. Suppose you want to look at a star. Choose one of the following stars:

Fall	*Spring*
Star: Fomalhaut	Star: Alpha Hydrae
RA: $22^h\ 58^m$	RA: $9^h\ 28^m$
Declination: $-29°\ 37'$	Declination: $-8°\ 40'$

    Show the location of this star on Figure 8.1.

2. Next plot the location of this same star as it would appear one hour later on Figure 8.2. This distance in hours, minutes, and seconds that a star is west of the central meridian at any given time is known as the hour angle of the object. Determine the hour angle of your star in Figure 8.1.

3. What is the hour angle of your star in Figure 8.2?

Exercise Eight: Observing with the Telescope, Part I

Figure 8.1
The sky looking south in spring and fall.

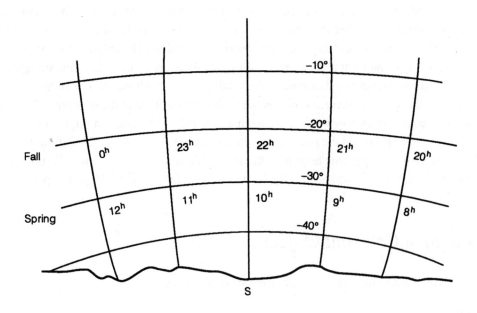

Figure 8.2
The southern sky one hour later.

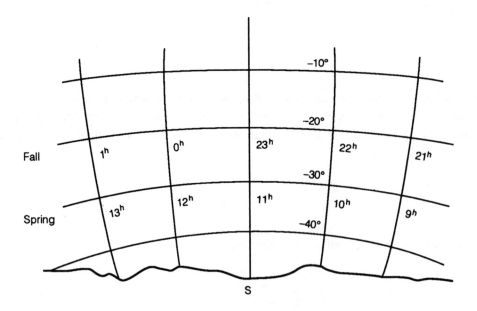

4. How much has the hour angle changed between the two diagrams?

5. Set up one of the large celestial globes in the lab so it corresponds to an observer at your location. This means that the north celestial pole should point toward the north and be elevated at an angle equal to your latitude from the horizontal plane. Determine the sidereal time from a sidereal clock. Next rotate the globe so that the central meridian of the globe has a right ascension under it that is equal to the sidereal time. Locate your star on the globe and notice its position in the sky. What are the altitude and azimuth of your star in the sky?

6. Go outside and locate your star in the sky using the coordinates you have found. Check with the instructor if you have been in error.

7. Return to the lab room and recheck the sidereal time. Record the present sidereal time below for reference.

8. Now subtract the right ascension of your star from the sidereal time. This is the hour angle of your star (ST = HA + RA) and is a measure of how far west it is from the central meridian.

9. Is the hour angle you have determined positive or negative?

10. Is your star east or west of the central meridian?

Exercise Eight: Observing with the Telescope, Part I

11. Go to the roof again and point one of the telescopes due south. Set the declination circle. Next move the telescope a distance in right ascension equal to the hour angle you have calculated. Normally this can be done by setting the right ascension to zero with the telescope pointed due south. The hour angle is then measured east or west from zero. With other telescopes the right ascension circle can be slipped so that the circle will read the sidereal time when the telescope is pointed south. In this case the object can then be found by merely turning the telescope to the right ascension of the object being sought. Try both methods if the telescope you are using has a slip-type right ascension circle. Why is it possible to leave the declination adjustment alone throughout this entire procedure?

12. An open star cluster is located at the following coordinates:

$$RA = 03^h\ 46^m \qquad Declination = +24°\ 10'$$

Find this cluster using the techniques that you have just used. If this cluster is not above your horizon, substitute one at $RA = 02^h\ 20^m$, declination $= +57°\ 15'$.

13. Draw in the bright stars of the cluster below as they appear in the telescope finder. On the diagram also indicate the approximate field of view of the main telescope and mark the directions of north and west. (North is the direction the stars appear to move in the eyepiece when the declination setting of the telescope is made smaller. West is the direction the stars move when the telescope drive is momentarily shut off.)

14. Look at Figures 8.3 and 8.4 and find the cluster using the known coordinates. In what constellation is the cluster located?

15. What is the field of view of the main telescope as indicated by comparing the stars included in your diagram to the separation of these same stars on the chart? (If west on your diagram is to the left when north is at the top, you should look at it in a mirror when comparing it to the charts.)

16. Some of the brighter stars in the cluster may be numbered on Figures 8.3 and 8.4. Identify some of these on your diagram. Note that in finding a faint object, you will usually find the coordinates of the object in a table listing many objects of the same type. The telescope is then moved to the proper field in the sky using right ascension and declination. Finally, a finder chart prepared in advance from a star chart must be used to locate the object in the field of stars visible through the telescope finder. This procedure is necessary even with the largest observatory telescopes. It is especially important when the object is invisible to the eye through the telescope and can therefore only be located by reference to stars that are visible.

Figure 8.3

Part of a skychart (from *Sky Atlas 2000.0*, reduced)

Source: Reproduced by permission. © 1981 Sky Publishing Corp.

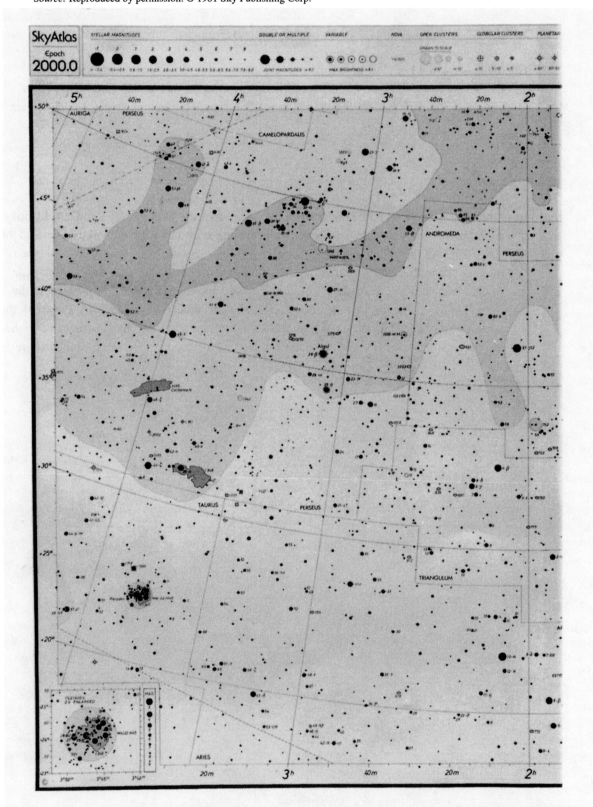

Figure 8.4

Part of a skychart (from *Sky Atlas 2000.0*, reduced)

Source: Reproduced by permission. © 1981 Sky Publishing Corp.

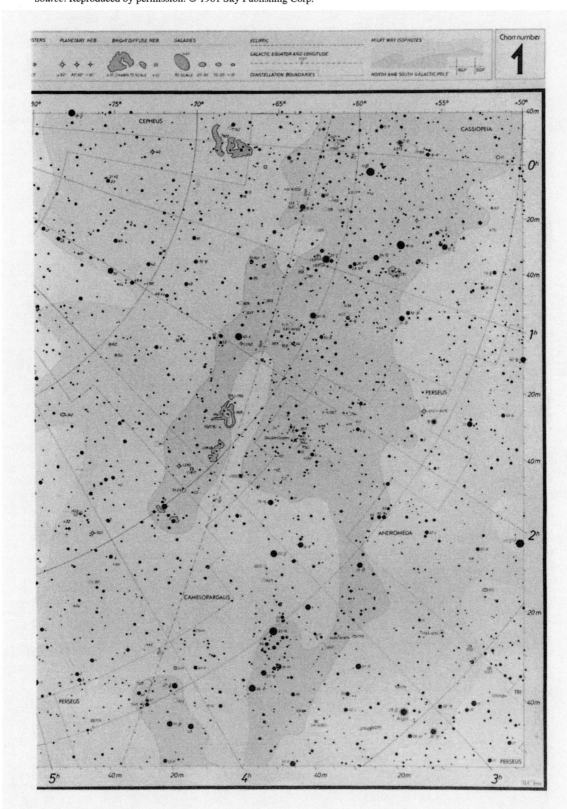

Name: _____

Exercise Ten: Outdoor

Observing with the Telescope, Part II: The Limitations of the Telescope

I. Light-Gathering Power

There are two properties of any astronomical telescope that are of importance to the astronomer. The first of these is the light-gathering power of the telescope, which depends most directly on the diameter of the telescope's objective lens or mirror. The unaided human eye collects an amount of light that is roughly equal to the surface area of the pupil (the circular aperture at the front of the eye). The only light the eye can detect is light that passes through this circular aperture, which is only about .20 inches in diameter. You can calculate the collecting area of the eye by knowing the formula for the area of a circle and using .20 inches for the diameter of the pupil:

$$Area = (\pi d^2)/4 = [3.14(.20)^2]/4 = .0314 \text{ inches}^2$$

The telescope, however, can collect light that falls over a much larger surface area than the pupil of the eye. The area of the objective divided by the area of the eye pupil is the light-gathering power of the telescope. A 2-inch diameter binocular lens, for example, has an area of 3.14 square inches. It therefore reveals objects that are 100 times fainter than those that are at the limit of the unaided eye. This corresponds to exactly five magnitudes of brightness. [Each step of one magnitude amounts to a factor of 2.512 in intensity. A total of five magnitudes therefore means a total change in intensity of $(2.512)^5 = 100$.] The normal naked-eye limit is about sixth magnitude. One should therefore be able to reach 11th magnitude with a good pair of binoculars having 2-inch diameter objective lenses.

1. What is the light-gathering power of a 20-inch telescope?

2. To the nearest whole magnitude, what are the faintest stars that can be seen with a 20-inch telescope?

3. Determine the limiting magnitude of the Palomar Mountain 200-inch telescope without actually calculating the collecting area of the primary mirror. Use information that you already have at hand from earlier parts of the lab.

4. What are some factors in telescope design and construction that might affect the amount of light that actually reaches the focus of an astronomical telescope?

II. Resolving Power

The second important characteristic of an astronomical telescope is its ability to show small detail. This ability is usually indicated by the telescope's resolving power, which is the angular separation of two stars that can just barely be seen as separate stars through the telescope under ideal conditions. A large telescope objective is capable of separating very close stars and therefore has a great resolving power. The limit of resolution for any telescope is usually given by Dawe's limit, which is a simple relation that shows how the objective diameter D determines the resolution limit of the telescope.

Angular resolution $= 4.5"/D$ (inches)

A 4.5-inch telescope will therefore resolve to 1 second of arc, a 9-inch telescope to .5 second of arc, and so on.

1. What is the smallest telescope that can be used to resolve the double star μ Draconis, which has a separation of 2.2 seconds of arc?

III. Focal Length and Magnification

Astronomical objects are so far away that light reaching the Earth from them is very nearly parallel. An exact image of the sky is created inside the telescope because the objective collects light from each point in the sky and directs it onto a particular point in the focal plane of the telescope. An eyepiece is then used as a magnifying glass to enlarge the image formed by the objective of the telescope. The total angular magnification is easily calculated from the focal lengths of the telescope and the eyepiece being used with it:

Magnification = Focal length of objective/Focal length of eyepiece

A 3-inch telescope having a focal length of 1200 mm will therefore magnify 100× when used with an eyepiece that has a 12-mm focal length.

1. What magnification would you obtain with a 24-mm focal length eyepiece in this telescope?

2. If you wanted a telescope, would you be better off buying one with a large objective or one that would magnify a great number of times? Explain your choice.

Find a bright star in a telescope and remove the eyepiece. Look for the image of the star in the focal plane of the telescope by holding your eye about 6 inches behind the eyepiece holder of the instrument. Moving closer to the telescope, notice that the light of the star appears to fill the entire area of the objective. This shows that light from each point on the objective contributes to the final image. Next place a ground glass at about the same place the eyepiece was before it was removed from the telescope. Determine the focal point by locating the position where the circle of light on the ground glass is smallest. The ground glass is then located where the rays from the star converge to a point in the focal plane. The focal length of the telescope is then found for a simple telescope by measuring the distance from the objective to the ground glass. For telescopes with curved secondary mirrors, a more complicated procedure is required.

Exercise Ten: Observing with the Telescope, Part II

3. Use the diagram to help measure the focal length of your telescope.

Objective   Ground glass

4. Using the measured focal length, calculate the magnifications of several of the eyepieces that are normally used with the instrument. Notice that the resolving power of the telescope does not depend on the magnification used with the telescope.

IV. Atmospheric Seeing

Certain atmospheric conditions place a limitation on the ultimate performance of an astronomical telescope. Clouds, wind, high humidity, and rapid temperature fluctuations may all adversely affect telescope performance. On days of high humidity water sometimes condenses out as dew on optical surfaces. Temperature fluctuations may cause the optics of the telescope to warp to such an extent that the imaging properties of the telescope are completely destroyed. Professional observatories are designed to minimize the effects of atmospheric phenomena, but even under optimal conditions you normally cannot reach the theoretical resolving power of a large telescope because of atmospheric seeing, which distorts the optical image.

The stars are so far away that they are really only points of light in the sky. The image you actually see through the telescope is really only a diffraction pattern created by the light waves from the star in passing through the circular aperture of the telescope. The diagram shows a perfect star image.

1. Look at a bright star near the zenith with *high magnification* and sketch the star as it really appears through the telescope.

2. Hold a hotplate in the air beyond the objective of the telescope and determine if the seeing conditions deteriorate. Does this experiment indicate that seeing is affected by conditions high in the atmosphere or by those close to the ground?

3. Finally, look through the telescope at a star that is close to the horizon. Is the seeing better or worse than it was at the zenith? Does the atmosphere do anything to disperse the colors of the star into a spectrum? Can you tell which color appears to be bent the most toward the ground as it passes through the atmosphere? If so, what is it?

1. Now perform the necessary measurements to fill in the table. Do all the calculations and scale drawings necessary (be neat). For your information, tan(5°) = 0.088 and tan(10.5°) = 0.19.

Figure	Sun's altitude	Width of crater (cm)	Feature	Length of shadow (cm)	Length of shadow (km)	Height (km)
11.17	5°		Tycho—rim			
11.17	5°		Tycho—peak			
11.15	10.5°		Archimedes—rim			

2. Using this information, make cross-section drawings of Tycho and Archimedes. You will see that it is necessary to exaggerate the vertical scale relative to the horizontal scale. Use an exaggeration of 10 times, and make the drawings on the graph paper provided at the end of the exercise.

3. What are the major differences between Tycho and Archimedes?

4. Suppose that after your measurement, Tycho filled up with lava 2 1/2 km deep, which then hardened. How would Tycho and Archimedes then be more similar?

5. Look at the crater Tsiolkovsky in Figure 11.7. Give a brief account of what you think happened to it after its initial formation.

Exercise Eleven: The Moon

6. Finally, look at Copernicus in Figure 11.18. Do you think it is younger or older than Archimedes? Why?

VI. Rilles

1. Lunar maria are often seen to be crossed by narrow linear depressions called *rilles* (see Figure 11.14). What Earth features do these remind you of?

2. Many scientists believe, from the evidence of moon samples, that the lunar lava was very thin and free flowing. What does this suggest about the origin of some of these rilles?

VII. Relative Crater Ages

1. The moon's surface is constantly being eroded away by countless impacts of microscopic meteoroids. Thus, although young craters formed by large impacts may be fresh with sharp edges, old craters have more rounded edges. From Figure 11.12, which crater would you say was older, Theophilus or Cyrillus? Why?

2. Now took at Tycho (Figure 11.17) and Copernicus (Figure 11.18). Are they young or old craters?

Exercise Eleven: The Moon

3. On the full moon photograph (Figure 11.11), the most pronounced aspect of the lunar surface near Tycho and Copernicus is the bright rays emanating from them. Scientists believe, on the basis of lunar missions, that the rays are caused by material thrown out of large craters when they are formed. Do the rays cast shadows?

4. What does this indicate about their nature?

5. Do all large craters have rays?

6. Judging by Copernicus and Tycho, would you say that craters with rays are young or old?

7. If all large craters had ray systems when they were formed, but only a few ray systems exist now, some of the ray systems must have been destroyed. What do you think could destroy the ray systems?

VIII. Crater Counting and the Highlands

Selenographers (those who study the moon) are interested in the past history of the moon's surface. If they know how the moon's surface has been changed, they can hope to find out what the original state of the moon was and what forces have been acting to change it. One way of determining this information is to investigate the relative numbers of small and large craters on different regions of the moon. This can be done by counting the number of craters of various sizes on a given part of the moon's surface. All craters within a given size range in a particular region are counted.

1. If you did a crater count on a highlands area and a similar count on a mare area, you would come up with fewer craters in each category in the mare area. Assume that the meteoroid impact rate has remained constant throughout the history of the moon (probably a bad assumption). If you counted 1/2 as many craters of each size in a mare area as in an equal area of highlands and the highlands are known to be 4.6 billion years old, how old are the maria?

2. If the impact rate has been diminishing with time, would this make the mare younger or older than your estimate in Question 1? Why?

Figure 11.2
Map of the moon's near side.
Source: U.S. Air Force Aeronautical Chart and Information Center.

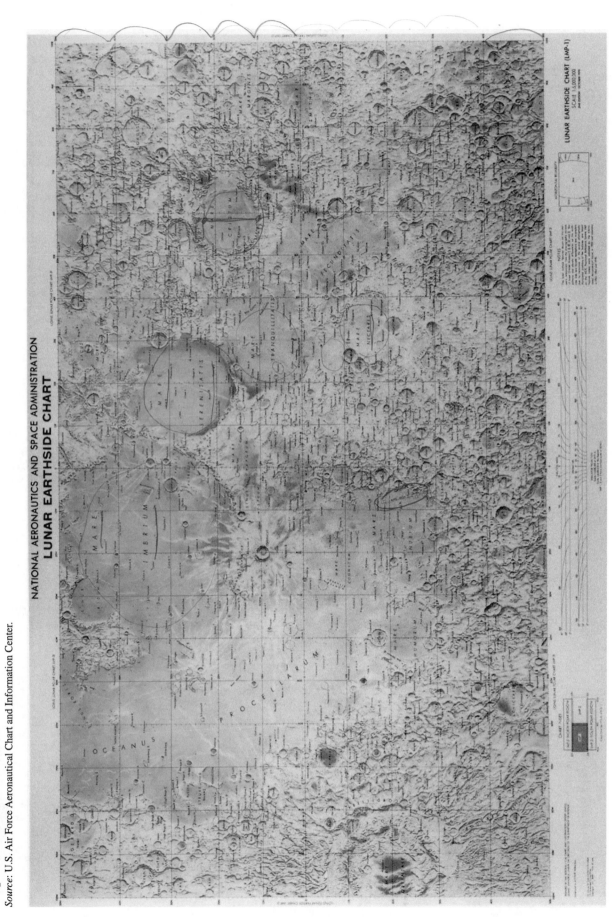

Figure 11.3
Map of the moon's far side.
Source: U.S. Air Force Aeronautical Chart and Information Center.

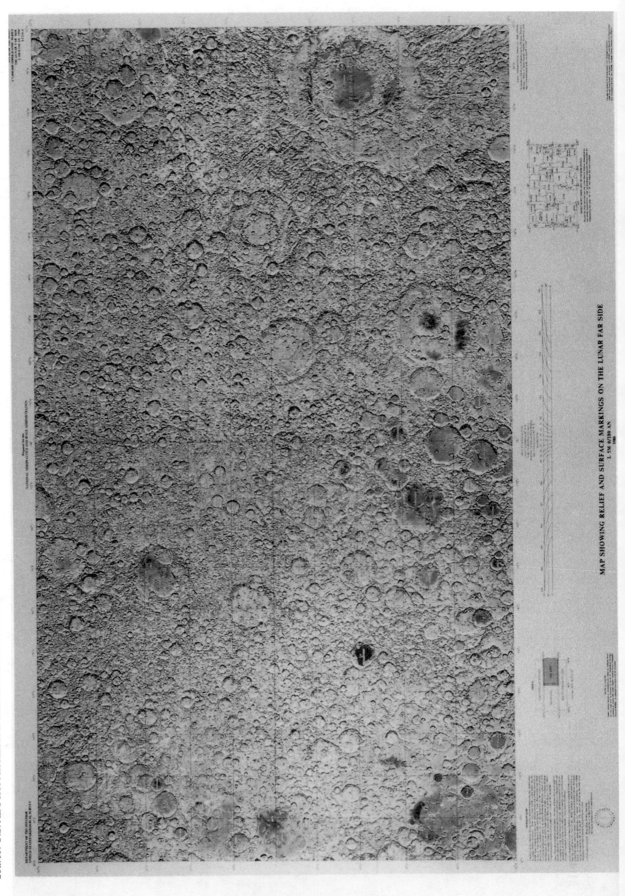

Figure 11.5
Last quarter moon.

Figure 11.4
First quarter moon.

Figure 11.6
Lunar Orbiter 4 photo of Mare Orientale from an altitude of 2700 km.
Source: NASA.

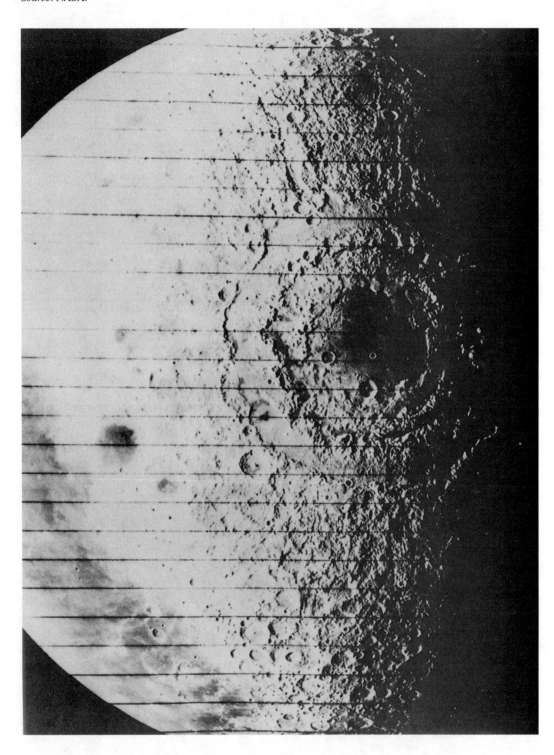

Exercise Eleven: The Moon

Figure 11.7
Lunar Orbiter 3 photo of the moon's far side from an altitude of 1500 km.
Source: NASA.

Exercise Eleven: The Moon

Figure 11.8
Lunar Orbiter 2 photo of the region of Copernicus.
Source: NASA.

Exercise Eleven: The Moon

Figure 11.9
Ranger 7 photo of the moon.
Source: NASA.

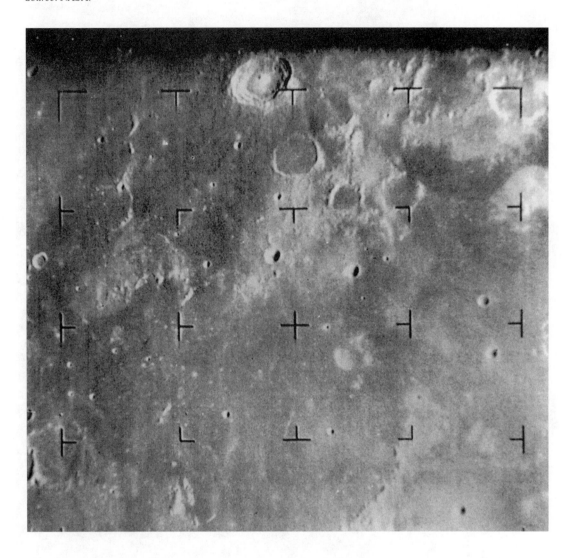

Exercise Eleven: The Moon

Figure 11.10
Mosaic of Surveyor 1 photographs taken in June 1966.
Source: NASA.

Figure 11.11

Full moon.

Source: G. P. Kuiper et al., *Photographic Lunar Atlas.* © 1960 by The University of Chicago. All rights reserved.

Exercise Eleven: The Moon

Figure 11.12
Theophilus (B5-a).
Source: G. P. Kuiper et al., *Photographic Lunar Atlas*. © 1960 by The University of Chicago. All rights reserved.

Exercise Eleven: The Moon

Figure 11.13

Fracastorius (B6-b).

Source: G. P. Kuiper et al., *Photographic Lunar Atlas.* © 1960 by The University of Chicago. All rights reserved.

Exercise Eleven: The Moon

Figure 11.14
Lunar rilles (C4-b).

Exercise Eleven: The Moon

Figure 11.15
Archimedes (D3-a).

Exercise Eleven: The Moon

Figure 11.16
Thebit (D6-b).

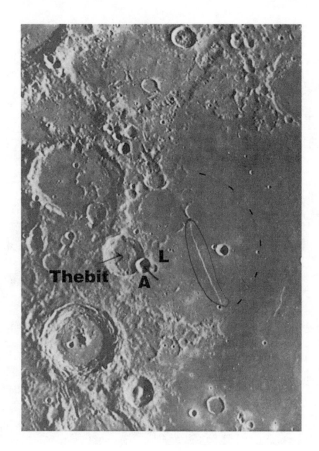

Figure 11.17
Tycho (D7-e).

Exercise Eleven: The Moon

Figure 11.18
Copernicus (E4-b).
Source: G. P. Kuiper et al., *Photographic Lunar Atlas.* © 1960 by The University of Chicago. All rights reserved.

Figure 11.19
Letronne (E5-a).

Exercise Eleven: The Moon

Name: _____

Exercise Twelve: Outdoor

Observing with the Telescope, Part III: Visual Observations of the Moon

I. Introduction

The visual observer has become virtually obsolete in modern astronomical research, but amateur and professional alike can still be thrilled by views of some of the many wonders in the night sky. Even a small telescope can perform superbly and give great satisfaction when used with a moderate degree of skill. As when learning any new skill, however, you should proceed with patience to avoid frustration. All telescopes are limited in capability, so do not be too disappointed when the views seen through a small telescope do not look anything like the impressive photographs that have been taken through the world's largest telescopes.

The moon is the nearest neighbor to the Earth in space and can be examined in detail with the aid of the astronomical telescope. The features that are visible depend somewhat on the phase of the moon. Answer only those questions that apply according to the moon's phase on your lab night.

II. Full Moon

1. At full phase the entire lunar hemisphere that faces the Earth is illuminated by the sun. Draw a diagram showing the relative position of the Earth, moon, and sun at full phase, and use this diagram to show why shadows on the Moon's surface would not be visible from the Earth. Verify this conclusion by observing the full moon through the telescope.

2. The full moon is conspicuously marked by long bright streaks radiating from several large craters on the lunar surface. Locate several of these streaks, called lunar rays, and determine the craters from which they appear to radiate. What are the names of these craters?

3. Look at the full moon without using the telescope. Notice that conspicuous light and dark areas are visible. Now using a map of the moon and the telescope, try to name as many of these areas as you can, especially the dark areas. (Be careful; some telescopes invert the images.)

III. Partial Phases

1. At times other than full phase, the Earth, moon, and sun do not lie along the same line. You therefore can see only part of the lunar hemisphere that is illuminated by the sun. Draw a diagram showing the relative positions of the Earth, moon, and sun on the day of the lab.

2. At times of partial phase, you can see the long shadows cast by the mountains and craters that lie close to the terminator, which is the dividing line between night and day on the lunar surface. With the aid of the lunar charts, locate several craters or other features that lie along the terminator. Make a sketch in the space provided on the next page.

3. Look at the mountains along the terminator under high magnification, and notice that the edge of the illuminated area is irregular because of the terrain. Look for mountain peaks that are just barely illuminated by the sun. Next examine the flat maria, and watch for ridges that may have been caused by lava flows. One prominent area of such ridges is in Mare Imbrium near Sinus Iridum. Can you explain why Sinus Iridum is not completely circular like most other features of the lunar surface? If Sinus Iridum is not visible, look instead at the crater Fracastorius.

4. Look at the craters Copernicus, Ptolemaeus, Plato, and Theophilus if they are visible. Which of them have central mountain peaks?

5. Do the smallest craters visible on the lunar surface have central peaks?

6. Look at the lunar fault called the straight wall. Is the east side higher or lower than the west side? How did you determine this?

7. Look for the smallest crater that can be clearly seen with your particular telescope. Estimate the size of your crater by comparing its width on a lunar chart to the diameter of the entire moon as shown on the same chart. The diameter of the moon on a lunar chart represents a true distance of 2160 miles.

Exercise Twelve: Observing with the Telescope, Part III

8. Lunar rilles are shallow depressions that run for long distances across the lunar surface. The most prominent of these is associated with the crater Hyginus. Locate it if possible. Look also for narrow valleys in the mountains surrounding Mare Imbrium. Several of these appear to radiate from a point at the center of Mare Imbrium itself. From the appearance of these rilles and valleys, can you suggest how these features might have been formed?

Name: _____

Exercise Thirteen: Indoor

The Planets, Part I: Analysis of Observations

The nine planets of our solar system are usually divided into two groups; the terrestrial planets and the Jovian planets. The terrestrial planets (Mercury, Venus, Earth, Mars, and Pluto) are those most like the Earth; they are relatively small, dense planets composed mostly of rocky and metallic material. The Jovian planets (Jupiter, Saturn, Uranus, and Neptune) are larger and of rather low density. (In fact, Saturn would float if there were a large enough pool of water.) The Jovian planets are probably composed mostly of hydrogen and helium. Planets closer to the Sun than Earth is (the inferior planets) show phases in a telescope. Superior planets don't.

I. The Terrestrial Planets

1. Take a look at several photographs and drawings of Venus (Figures 13.1 through 13.4). What do you first notice about the changing appearance of this planet—that is, something analogous to the changing appearance of the moon? How are Mercury (Figure 13.21) and Mars similar or different in this respect?

2. The amount of energy a planet receives from the sun is inversely proportional to the square of its distance from the sun. What effect might this have on the relative surface properties of the terrestrial planets? In what way does this significantly affect you?

3. What observations do you think have been made to establish that Venus has an appreciable atmosphere? Again, take a look at Venus in Figures 13.1 through 13.4.

4. a. Examine several of the Earth-based photographs of Mars (Figures 13.5 and 13.6). What sort of general features do you observe? Do you notice any evidence for seasonal variations on Mars? If so, what?

   b. Now examine the photograph taken by the Viking Orbiter (Figure 13.7) and the Mars maps prepared from Mariner 9 data (Figures 13.8, 13.9, and 13.10). What sort of features do you observe? The Martian surface resembles the surfaces of which other celestial bodies? What do the Martian surface features tell us about the Martian atmosphere?

   c. What sort of earthbound observations do you think can be made to establish that Mars has any atmosphere at all?

II. The Jovian Planets

1. a. Examine the several photographs of Jupiter (Figures 13.11, 13.12, and 13.13). Do you think that we are looking at a solid surface of Jupiter or at the upper layers of its atmosphere? Explain. What kind of observations would help us to determine whether we are looking at atmospheric phenomena or at a solid surface?

   b. What do you notice about the shape of Jupiter? How might this shape be explained?

   c. Notice the bands (of different colors) running across the disk of Jupiter. How might these bands be related to the motion of Jupiter?

   d. On several of the photographs of Jupiter, you can see a boat-shaped feature (that is reddish in color). This is called the Great Red Spot. How might one use this spot to determine the sidereal rotation period of Jupiter?

   e. Figure 13.14 shows the remarkable telescopic sketches of the surfaces of the four Galilean satellites of Jupiter by Lyot (an astronomer at the French observatory Pic du Midi). Compare the Voyager maps of the Galilean moons (Figures 13.15, 13.16, and 13.17) to the Lyot sketches. On the rough scale of Lyot's sketches, are they similar or different? Could telescopic sketches be of use in evaluating changes on Io?

2. a. Saturn, which is the sixth planet from the sun, has a beautiful set of rings around it (Figure 13.18). Can you suggest how these rings might have been formed? (Consider gravitational tidal forces.)

   b. Between Saturn's outer two rings is a gap known as Cassmi's division (Figures 13.19 and 13.20). Can you think of a mechanism by which this gap was created? (Again, consider gravitational forces.)

   c. Do you think that Saturn's composition and internal structure are similar to or different from Jupiter's? Why?

3. Unlike other planets, Uranus's axis of rotation lies almost in the plane of its orbit. How would this affect its seasons and the length of its days (or nights)?

III. The General Properties of the Solar System

1. Why do you think that the planets are all nearly spherical in shape and have very dense cores? (Consider gravity.)

2. All the planets revolve about the sun in a counterclockwise direction as viewed from the north, and the sun and all the planets (except Venus, Uranus, and Pluto) have axial rotations that are counterclockwise. Also, most of the natural satellites revolve about their respective planets in this same direction. The orbital planes of all of the planets lie nearly in the ecliptic plane, and, except for Uranus and Pluto, the equatorial planes of the planets are inclined less than 30° to their orbital planes. How would you explain this regular behavior of the solar system objects?

3. In a gas the following relation holds: $V^2 = 3kT/m$, where V^2 is the square of the average speed of the molecules, T is the temperature of the gas, m is the mass of one molecule, and k is a constant. For a planet, the escape velocity is given by the formula $V_{escape}^2 = 2GM/R$, where G is the gravitational constant, M is the mass of the planet, and R is the planet's radius. Use these two equations to explain why the atmospheres of different planets have different compositions.

Exercise Thirteen: The Planets, Part I

Figure 13.1

Photos of Venus taken on different dates with a 15-inch refractor.

Source: G. P. Kuiper and B. M. Middlehurst, *Planets and Satellites*, Chicago; University of Chicago Press, 1961.

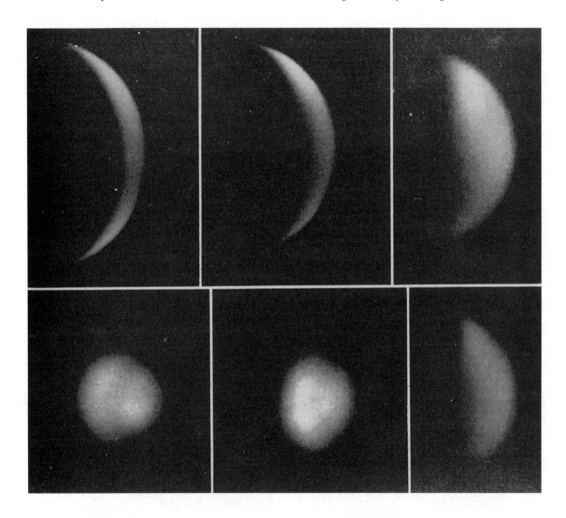

Figure 13.2
The crescent Venus.
Source: G. P. Kuiper and B. M. Middlehurst, *Planets and Satellites*, Chicago; University of Chicago Press, 1961.

Figure 13.3
Drawings of Venus as observed when transiting the face of the sun.
Source: L. Rudaux and G. de Vaucouleurs, *The Larousse Encyclopedia of Astronomy*, 2nd ed., New York, NAL Books, 1962, p. 190.

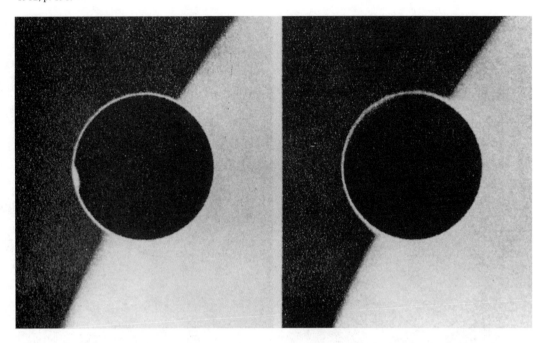

Figure 13.4
Venus's crescent elongated by its atmosphere.
Source: G. P. Kuiper and B. M. Middlehurst, *Planets and Satellites*, Chicago; University of Chicago Press, 1961.

Exercise Thirteen: The Planets, Part I

Figure 13.5
Excellent earthbound photos of the 1971 opposition of Mars.
Source: R. Minton and S. Larson, Lunar and Planetary Laboratory, University of Arizona.

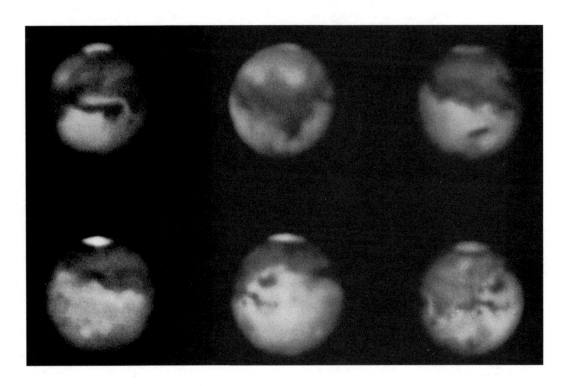

Figure 13.6
Seasonal changes in the polar caps and dark markings on Mars.
Source: Lowell Observatory.

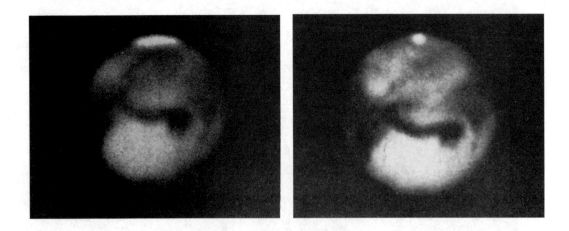

Figure 13.7
Viking Orbiter photos of channels on Mars, apparently cut by running water.
Source: NASA.

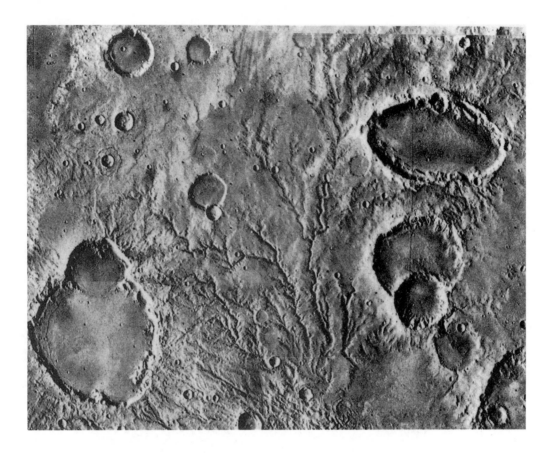

Figure 13.8
Mariner 9 map of Mars, with albedo from Earth-based photography.
Source: Lowell Observatory.

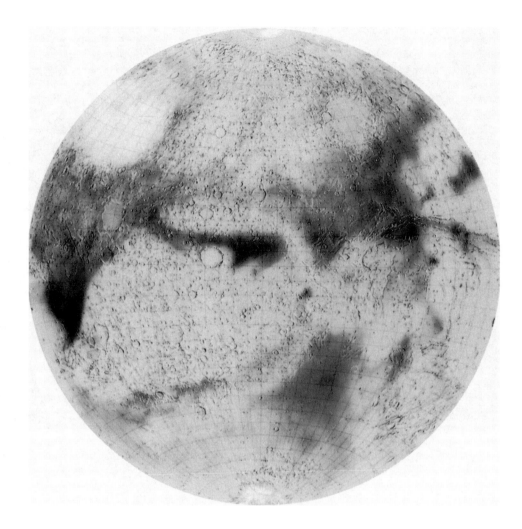

Figure 13.9
Mariner 9 map of Mars, second view.
Source: Lowell Observatory.

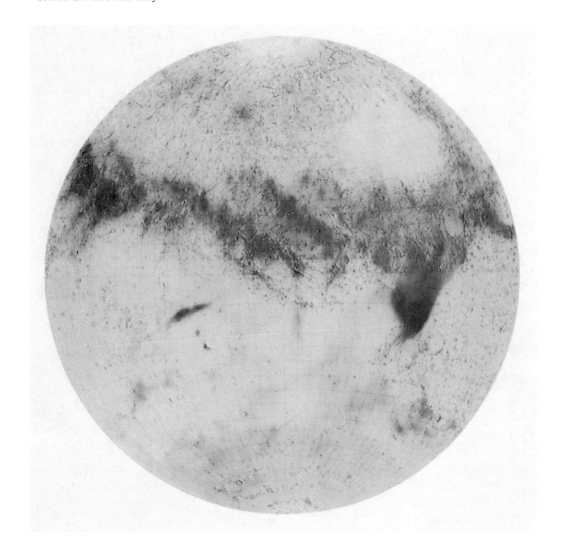

Figure 13.10
Mariner 9 map of Mars, third view.
Source: Lowell Observatory.

Figure 13.11
Pioneer 10 photos of Jupiter being crossed by the shadow of Io.
Source: NASA.

Figure 13.12
Jupiter's Great Red Spot, imaged by a Voyager spacecraft.
Source: NASA.

Figure 13.13
Voyager 1 photo of Jupiter, taken from a distance of 3×10^7 km.
Source: NASA.

Figure 13.14
Earthbound maps of Jupiter's Galilean moons, drawn by Lyot.
Source: G. P. Kuiper and B. M. Middlehurst, *Planets and Satellites*, Chicago; University of Chicago Press, 1961.

Io

Europa

Ganymede

Callisto

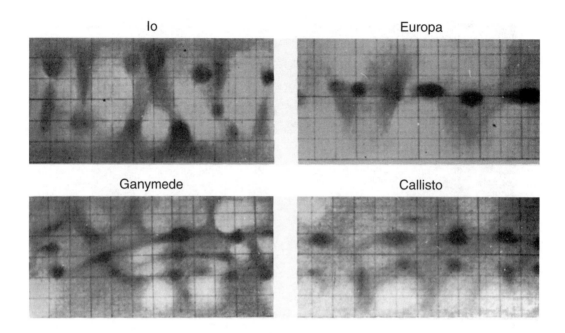

Figure 13.15
Voyager map of Io.
Source: NASA.

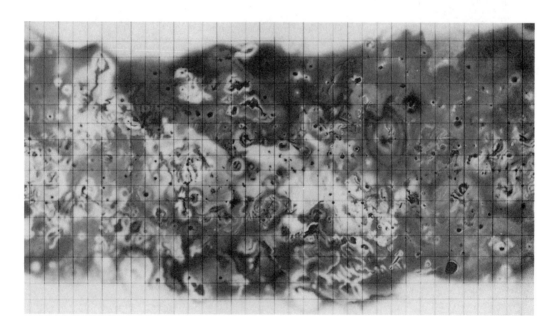

Exercise Thirteen: The Planets, Part I

Figure 13.16
Voyager map of Europa.
Source: NASA.

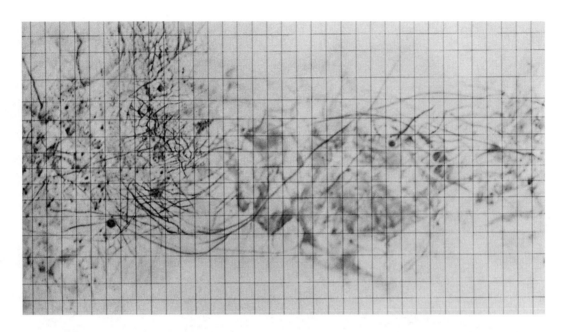

Figure 13.17
Voyager map of Ganymede.
Source: NASA.

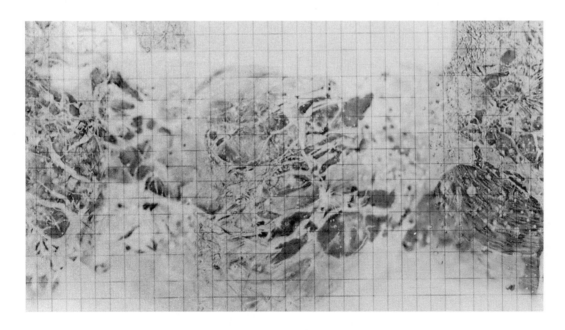

Exercise Thirteen: The Planets, Part I

Figure 13.18
Earthbound photo of Saturn in blue light.
Source: G. P. Kuiper and B. M. Middlehurst, *Planets and Satellites*, Chicago; University of Chicago Press, 1961.

Figure 13.19
Telescopic aspect of Saturn's rings, drawn by Lyot under conditions of excellent seeing.
Source: G. P. Kuiper and B. M. Middlehurst, *Planets and Satellites*, Chicago; University of Chicago Press, 1961.

Exercise Thirteen: The Planets, Part I

Figure 13.20

Voyager photos of Saturn's rings.

The rings of Saturn change greatly in appearance under different lighting and viewing conditions. The view of the sunlit face, with the sun behind, is the most familiar (A). From below the rings, Voyager saw the unilluminated face (B), which shines faintly by diffusely transmitted sunlight. A third perspective was obtained after encounter (C), when the sunlit face was seen again, with the spacecraft looking back toward the source of illumination. *Source*: NASA.

a.

b.

c.

Figure 13.21
Mariner 10 photo of Mercury.
Source: NASA.

Exercise Thirteen: The Planets, Part I

Figure 13.22

The motion of Pluto, photos taken one day apart with the McDonald Observatory 82-inch telescope.

Source: G. P. Kuiper and B. M. Middlehurst, *Planets and Satellites*, Chicago; University of Chicago Press, 1961.

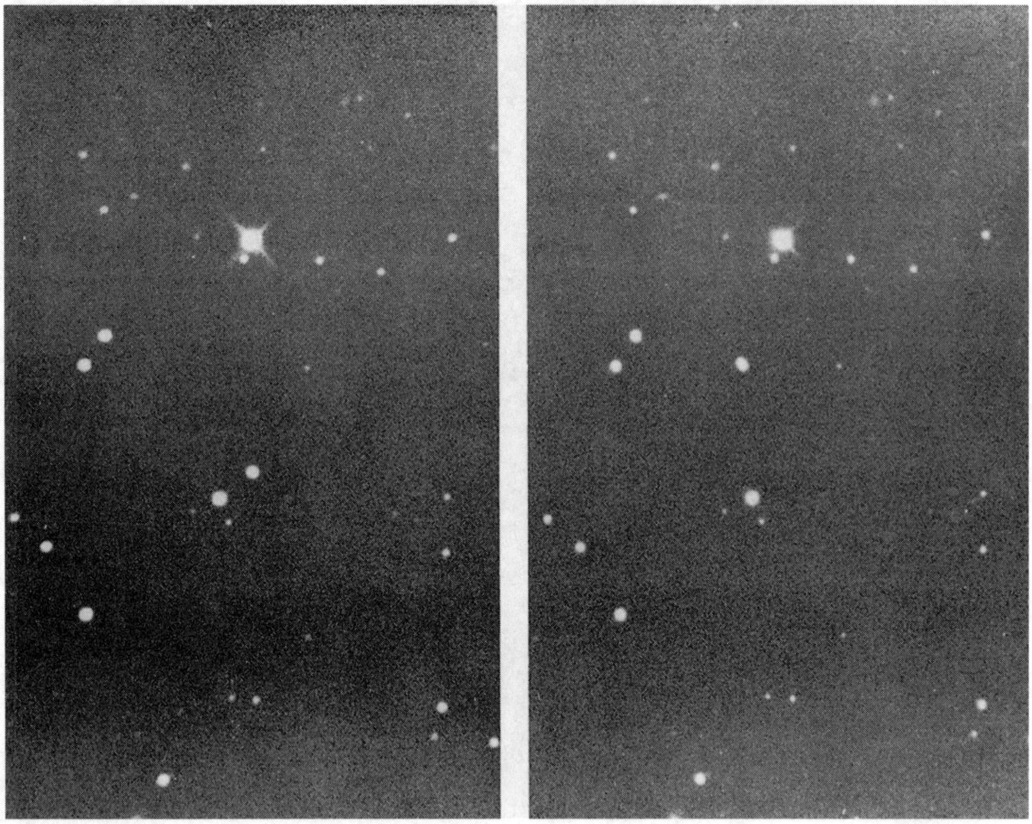

Name: _____

Exercise Fourteen: Outdoor

The Planets, Part II: Observations with the Telescope

Of the nine planets of our solar system, five are easily visible to the unaided eye and are easily located. These are Venus, Mars, Jupiter, Saturn, and, of course, Earth. Two others, Mercury and Uranus, can be seen with somewhat greater difficulty. Mercury is usually bright but is always very near the sun, so an observer must know exactly where to look right after sunset or right before sunrise. To find Uranus you need some sort of sky chart and binoculars or a small telescope. Neptune and Pluto are always rather faint, so larger telescopes and greater observing skills are needed to locate these two.

In this exercise you will be looking at those planets that are visible on the night of the lab and trying to observe some of their more noticeable features.

I. Locating the Planets

1. Look up the positions of Venus, Mars, Jupiter, and Saturn in the *Observer's Handbook* (published annually by the Canadian Royal Astronomical Society) or in *The Astronomical Almanac* (U.S. Naval Observatory). For Mars, Jupiter, and Saturn, look up their right ascensions and declinations, and then locate their positions on the celestial globe. What constellations are they near? Which ones are now visible? For Venus, look up its elongation (the angle between it and the sun). Is Venus visible tonight?

2. Now go to the roof and, using a telescope, find the planets that you can.

II. Angular Sizes of the Planets

The following equation can be used to determine the angular size of a planet or the field of view of a telescope:

Angular size = $15t \cos(d)$

where the angular size is measured in seconds of arc, t is the time in seconds required for the planet to drift past a point fixed with respect to the observer or for a star to drift across the field of view of a fixed telescope, and $\cos(d)$ is the cosine of the declination (this will be given to you by the instructor).

This equation can be used in two different ways to determine the angular size of a planet. One method is to find how long it takes for a star to drift across the field of view of a telescope (with the clock drive turned off). This time, t, can then be put into the equation to determine the field of view of the telescope. You then estimate the planet's angular size by estimating what fraction of the field of view it covers.

For the second method, you place the planet just inside the field of view of the telescope and then note how long it takes for it to drift out of sight. (Again, this should be done with the clock drive turned off.) Putting this time into the equation gives the angular size directly.

Exercise Fourteen: The Planets, Part II

1. How is the angular size of a planet related to its actual diameter? Suppose you determine that two planets have the same angular size, but that one is 10 times as far from the Earth as the other. What does this tell you about the relative sizes of the two planets?

III. Visual Observations

Answer only questions about the planets you can observe on the lab night.

1. Venus: Measure its angular size. What is most striking about its telescopic appearance? Do you see any surface features? What evidence, if any, can you detect that suggests Venus has an appreciable atmosphere? Sketch what you see.

2. Mars: Measure its angular size. What is the color of the planet? What surface features can you see? Can you see the polar caps, the dark regions of grayish or greenish shade, or the "canals"? Sketch what you see.

3. Jupiter: Measure its angular size. Does Jupiter appear spherical or does one of its diameters appear to be larger than another? Can you see any of its moons? Are the shadows of any of the moons visible on the surface of the planet? What surface features can you see—bands parallel to the equator or the Great Red Spot? What are the colors of the features? Sketch what you see.

4. Saturn: Measure the angular size of the planet and of the major and minor axes of the ring system. Can you see any of Saturn's moons? What surface features can you see? Are any of the divisions of the ring visible? How would you go about determining the inclination angle of the ring system? Do so. Knowing that Saturn's diameter is around 72,000 miles, determine the diameter of the ring system. Sketch what you see.

Exercise Fourteen: The Planets, Part II

5. Mercury and/or Uranus: If you or the instructor is able to find one of these planets, describe and sketch what you see. Does Uranus show any color? If so, what color is it?

6. Neptune and Pluto: Good luck!

Name:

Exercise Fifteen: Outdoor (Daytime)

Observing with the Telescope, Part IV: Visual Observations of the Sun

The sun is too bright for us to see surface features with the naked eye. It is safe, however, to view the sun through special filters placed over a telescope's objective (objective filters), through the combination of a dense filter and a Herschel wedge, which reflects only a small fraction of the sunlight into the eyepiece, or by projecting its image onto a white screen. *Never* look at the sun through an unfiltered telescope. This lab will acquaint you with the various aspects of the sun's surface as seen in visible light.

I. Solar Features

Using a low-power eyepiece, look at the entire sun. (It may be colored by the filters used to block out much of the sunlight.) Concentrate on trying to see the following features:

- *Limb darkening*: nonuniform brightness of the solar image.
- *Sunspots*: dark spots that move with the solar image when the sun is moved about in the telescope field. (Other spots are probably specks of dust on the telescope optics.)
- *Faculae:* light, patchy areas usually most visible near the edge of the sun.
- *Granulation:* fine structure (grainy appearance) all over the rest of the sun's disk. This is usually visible only when seeing conditions are very good.

Exercise Fifteen: Observing with the Telescope, Part IV

1. Quickly sketch the entire solar disk, identifying approximately where these features are most noticeable.

2. Why does the sun appear darker at the edge than at the center? In other words, explain limb darkening. (Hint—hotter areas appear brighter, and the deeper one can see into the sun, the hotter it is.)

3. Draw a close-up of a large sunspot (or group) as seen through a high-power eyepiece. Look for a group featuring two very close, large spots or for a large single spot.

4. Why do large spots occur in pairs or singly with a light "bridge" across the middle? (Hint—consider the sun's magnetic field.)

II. Solar Rotation

If you watch a particular sunspot group for a number of days, you see that the group moves across the disk of the sun. This motion indicates that the sun is rotating. By following the motion of sunspot, you can determine the rotation period of the sun. However, there is a complication, called the projection factor, which would cause you to overestimate the sun's rotation period if you do not use the correct procedure. Figure 15.1 is the polar view of the sun showing the northern hemisphere of the sun divided up into 10° segments of longitude. (In the figure the Earth would be in the direction of the bottom of the page.)

Figure 15.3 is an equatorial view of the sun. It is divided up into 10° segments both in apparent longitude and in latitude. (Notice that the 45° nearest the edge appears only 1/3 of the way from the solar rim; this is an example of the projection factor.) By using Figures 15.1, 15.2, and 15.3 you can compensate for the projection factor and obtain a more accurate value for the sun's rotation period.

1. On top of Figure 15.2, draw as carefully as you can the positions of as many spots as possible. You should locate at least six groups, three of which are at different latitudes.

2. After at least two days have passed, reobserve the sunspots and plot their new positions using a different color to distinguish these from the other positions. Be sure to use the same orientation of Figure 15.2 when plotting the positions for the two different days. (Why?)

3. Now, overlay Figure 15.3 with Figure 15.2, aligning the lines of solar latitude with the apparent movements of the spots on Figure 15.3.

4. Record in a table the two positions (longitude) and times of observation for each spot.

5. For each spot calculate the time necessary for the spot to rotate 360°. Average the values you get from spots at about the same latitude. Keep separate the values for spots at different latitudes. (How do your results compare with the accepted values, 25 days at the solar equator and 34 days at the poles?)

6. Plot a graph of spot rotation period versus solar latitude, using the graph paper provided at the end of the exercise.

7. What does this graph tell you about the surface of the sun?

Exercise Fifteen: Observing with the Telescope, Part IV

III. The Solar Spectrum

1. Using a hand-held spectroscope, look at a white card held in bright sunlight, and draw in the observable features of the solar spectrum in the space provided.

Blue
end

Red
end

2. What information about the sun does such a spectrum give us?

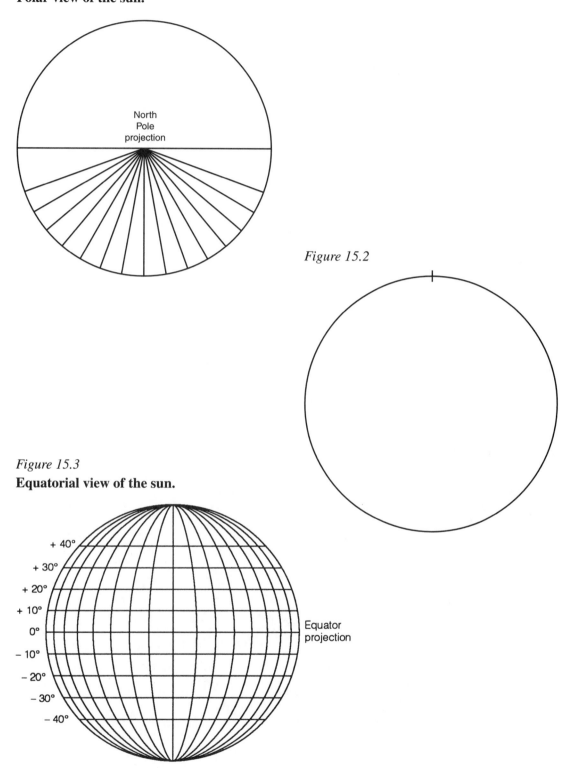

Figure 15.1
Polar view of the sun.

North
Pole
projection

Figure 15.2

Figure 15.3
Equatorial view of the sun.

+ 40°
+ 30°
+ 20°
+ 10°
0°
− 10°
− 20°
− 30°
− 40°

Equator
projection

Exercise Fifteen: Observing with the Telescope, Part IV

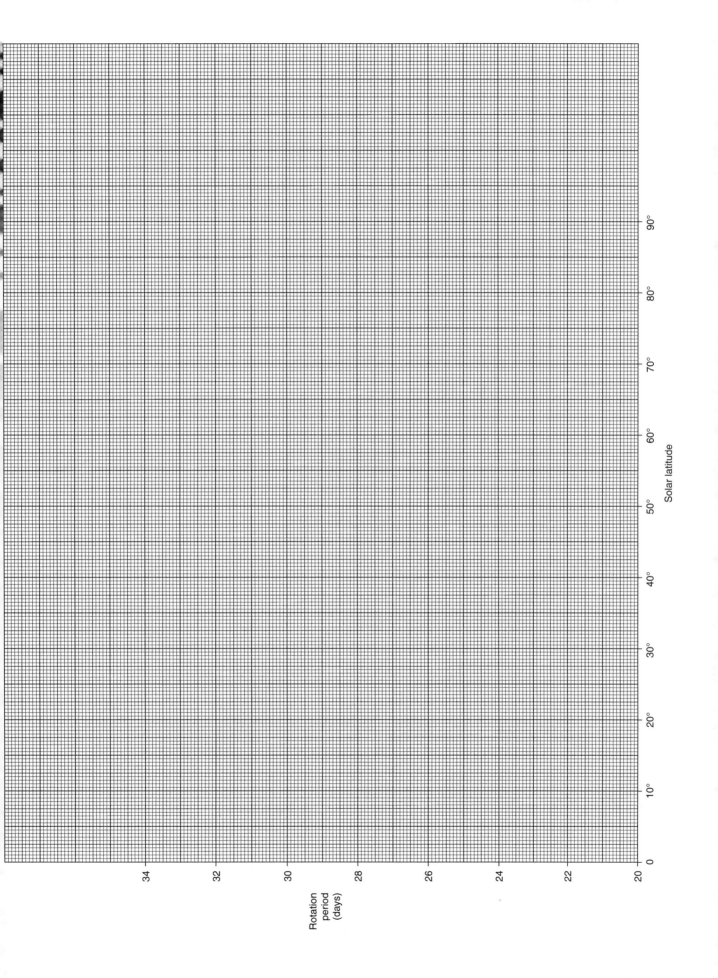

Solar latitude

Rotation period (days)

Name: _____

Exercise Sixteen: Outdoor

Measurement of Astronomical Distances

I. Introduction

One quantity of a body being observed that the astronomer needs to determine is its *distance*. This is necessary in order to learn the true characteristics (size, brightness, relative position, and motions) of celestial objects such as planets, stars, nebulae, and galaxies. The method used in this lab to find distances is similar to that used by land surveyors.

II. Parallax

If you hold a pencil up before your face (about 12 to 18 inches away) and look at it alternately with only the right and then only left eye open, the pencil appears to shift back and forth with respect to more distant objects (such as a wall). The pencil itself does not move; it merely *appears* to shift because it is viewed from two slightly different positions, separated by the distance between your eyes.

The position of the pencil with respect to the background (wall) seems to change because of the position from which it is viewed (right or left eye). This apparent change of position is called parallax. If the pencil is held farther away, the parallax decreases. Therefore there is some relation between the distance of the object and the amount of parallax we observe.

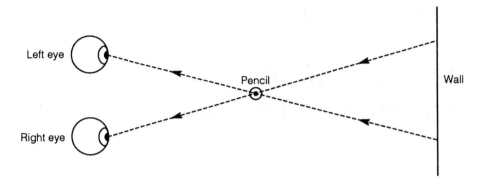

III. Survey Techniques

The fundamental way to determine the distance to an object by surveying techniques is one of triangulation. For example, if you want to measure the distance across a river without getting wet follow this procedure.

Select an object on the opposite bank—for example, a tree. Opposite this tree, mark off a straight line parallel to the river. This is called the baseline. At each end of the baseline, measure the angle as shown by sighting to the tree. Then, knowing all three angles of the triangle (the sum of the three angles is 180°) and the length of the baseline, it is possible to calculate the distance to the tree, that is, the line *AT* or *BT*. This is most easily done if the triangle used is a right triangle (*AOT*). The distance to the tree, *OT*, can be found if the angle *a* and the line *AO* are known.

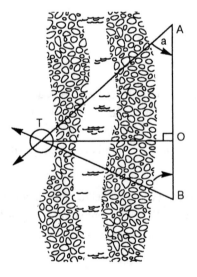

Exercise Sixteen: Measurement of Astronomical Distances

The distance to objects can be measured accurately by this technique to several times the distance of the baseline. It is clear that the longer the baseline, the farther away an object can be measured.

IV. Triangulation in Astronomy

Because the stars are so far away, it is impossible to get a long enough baseline on the Earth to measure these distances accurately. A much longer baseline is needed. One possible baseline is the diameter of the Earth's orbit around the sun: The "nearby" star is viewed at six-month intervals from opposite sides of the sun. Such a star would appear to move back and forth across the background of much more distant stars.

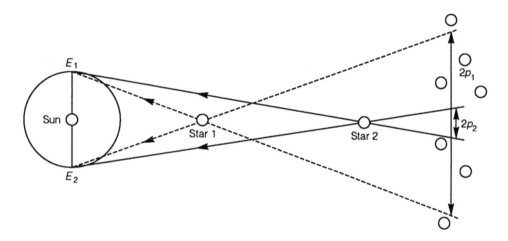

A photograph taken of this star field at E_1 and compared with one taken six months later at E_2 would show that the "nearby" star has moved across the background. The amount of this motion across the background depends on how close the star is: The farther away a star, the less will be the apparent shift back and forth.

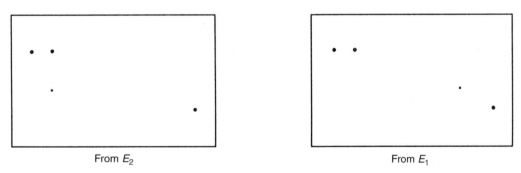

Exercise Sixteen: Measurement of Astronomical Distances

The parallax angle or parallax of a star in astronomy is defined as being equal to one-half the back-and-forth oscillation (over a six-month period) across the background. Measured in degrees, this angle, called p, is one of the angles of a right triangle.

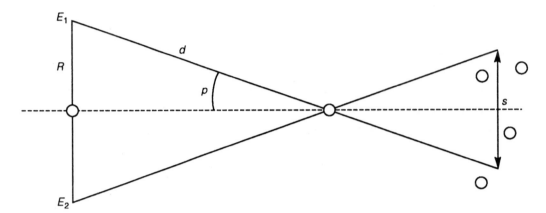

On two superposed photographs taken at these times, E_1 and E_2, the amount of oscillation is s. The parallax (or parallax angle) for this star

$p = s/2$

if s is measured in angular units.

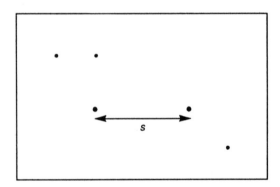

The result of determining the parallax of a star using the radius R (or diameter = $2R$) of the Earth's orbit around the sun is called heliocentric parallax.

The distance d from the Earth to the star is given by a trigonometric relation for right triangles. (Trigonometry is a specialized version of geometry.) If the angle p is small—that is, if the oscillation on the photograph is small (which is true for stars)—then the relation is

$d = R/p$

where R is the distance from the Earth to the sun, p is the parallax of the star, and d is the distance to the star.

Exercise Sixteen: Measurement of Astronomical Distances

V. Measurements of Heliocentric Parallax

This method, which is actually employed in astronomy, is extremely accurate for some of the nearer stars. However, even the nearest star has a parallax of much less than 1 degree. The angle p is extremely small, even for the nearest star.

We can subdivide one degree of angular measurement. One degree is equal to 60 minutes of arc, and 1 minute of arc (or angle) is equal to 60 seconds of arc:

$1°$ = 60 minutes of arc = 60' arc

1' arc   = 60 seconds of arc = 60" arc

Therefore

1" arc= (1/3600) of $1°$

One second of arc is only a very small fraction of a degree. And it has been found that the nearest star to the sun has a parallax of just under 1" arc! The angle p is at most 1" arc for the *nearest* star.

The distance between the Earth and the sun is about 93 million miles. This distance is called 1 astronomical unit (1 AU).

Because of the great distances to stars, it becomes convenient to define a new unit of distance. According to the formula for distance,

$d = R/p$

If R is measured in astronomical units ($R = 1$ AU) and the parallax p is measured in seconds of arc, then the distance calculated in this way is in units of parsecs (pc). That is,

$d(\text{pc}) = 1(\text{AU})/p("\text{arc})$

In units of parsecs, therefore

$d = 1/p$

where the parallax angle p (one-half the oscillation across the background) is measured in seconds of arc.

VI. Laboratory Procedure

The experiment is divided into two parts. First you will measure the distance to two nearby lampposts (stars) using the parallax method. Then you will measure two copies of photographic plates to determine the distance to a few nearby stars.

Part 1: Parallax Method

Your instructor will lay out a baseline with one observation point in line with two lampposts as "nearby stars" (Figure 16.1), and the other point a few meters away perpendicular to the lamp line (Figure 16.2). A faraway landmark (such as a tower) represents a distant star. To determine the distance to the nearby stars, you will measure the angles A and B (Figure 16.2) in order to deduce angle C, which equals $B - A$. [This can be seen by constructing a line through the near star to the distant star, which will be essentially parallel to those through points N and S, the ends of the baseline (Figure 16.3), and then noting that angle $A' = A$ and $B' = B$ (internal angles of parallel lines) and hence $C = B' - A' = B - A$.] The distance b is measured with a meter stick. A scale model triangle $P'N'S'$ similar to the triangle PNS can now be constructed on graph paper. Hence the distance to the star d can be determined by measuring the scale model triangle and converting back to the actual distance.

1. To measure angles A and B you will use a sextant. Your instructor will demonstrate its use. Become familiar with using it in class, and then go outside and measure the angles A and B for both of the near stars indicated in Figure 16.1. Record your results here:

	Angle A	Angle B	Angle $C = B - A$
Star 1			
Star 2			

2. Your instructor will tell you the distance b in meters. Return to class and draw the angle C for each star with a protractor on the enclosed piece of graph paper, with one side of the angle overlapping the graph paper lines. Find the perpendicular to the line that has a length in centimeters equal to the baseline b in meters for star 1. Draw that line to complete the triangle $P'N'S'$ (Figure 16.4), which should be a scale model of the triangle PNS, one hundredth actual size. Now measure the length of d, the scaled distance to star 1 in centimeters, by counting graph paper squares; this should be one hundredth the actual distance to star 1. Repeat for star 2, but use a scale of 5 mm on graph paper = 1 m. Record the results here:

   $d_1 = $ _____ $d_2 = $ _____

3. Which distance is more accurate? Why?

Figure 16.1

Observation point in line with lampposts.

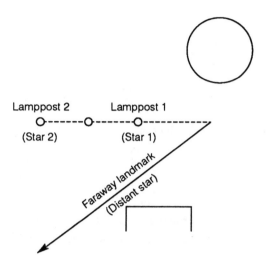

Figure 16.2

Baseline perpendicular to lamp line.

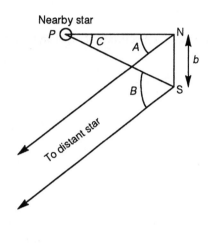

Figure 16.3

Line from distant star to nearby star.

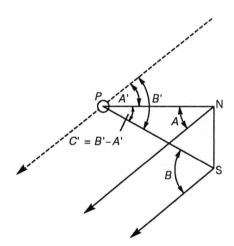

Figure 16.4

Completed triangle.

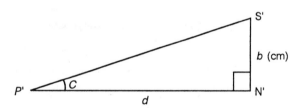

4. How accurate do you think your distances are?

5. What are the main sources of error in your measurement?

6. If you made an error of 1° in angle C, would it have a greater effect on star 1 or on star 2?

7. Find the distance error that a 1° angle error would cause by drawing angle C for star 2 one degree smaller and repeating the calculation of distance d. Does d increase or decrease? By how many meters?

Part 2: Distance to Nearby Stars

Included at the end of this lab are two diagrams that show several nearby stars against a field of very distant stars. A scale is included on the diagram in the form of a small line. The length of this line corresponds to 0.22" arc on this photograph.

1. Measure the above-mentioned line in millimeters and divide to find the scale, the number of seconds of arc per millimeter on the diagrams. Now you can convert any measurement in millimeters to seconds of arc. Record the result here:

   1 mm = _____ " arc

2. Superpose the two diagrams and determine the length of displacement of each of the five numbered stars in millimeters and record them in the data table. Convert into seconds of arc and enter into the table.

Star	Displacement (mm)	(")	Parallax	Distance (pc)	(ly)
1					
2					
3					
4					
5					

3. Recalling that the parallax angle p is one-half the six-month displacement, enter the parallax of each star and also the distance in parsecs (pc) to each from the formula

$d = 1/p$

where d is in parsecs and p is in arc seconds.

4. Finally, note there are 3.26 light years (ly) per parsec. Enter the distance to each star in light years into the data table.

5. The smallest parallax that can be measured accurately is 0.025" arc. What distance is this in parsecs? In light years?

6. While the star images on these diagrams look very small, on real astronomical photographs the displacement due to parallax might be much smaller than the size of the star images. Why?

Exercise Sixteen: Measurement of Astronomical Distances

3 Feb. 1997

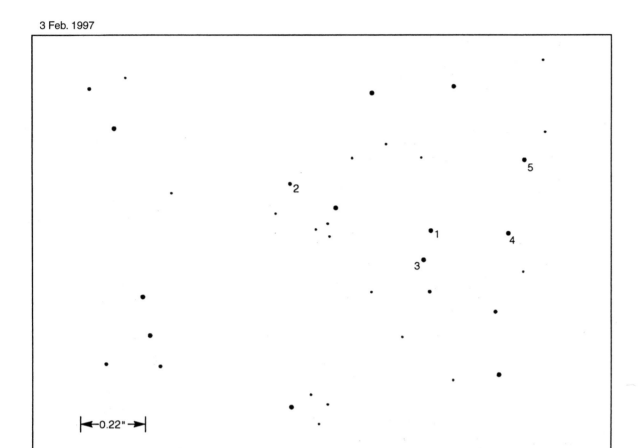

Exercise Sixteen: Measurement of Astronomical Distances

3 Aug. 1997

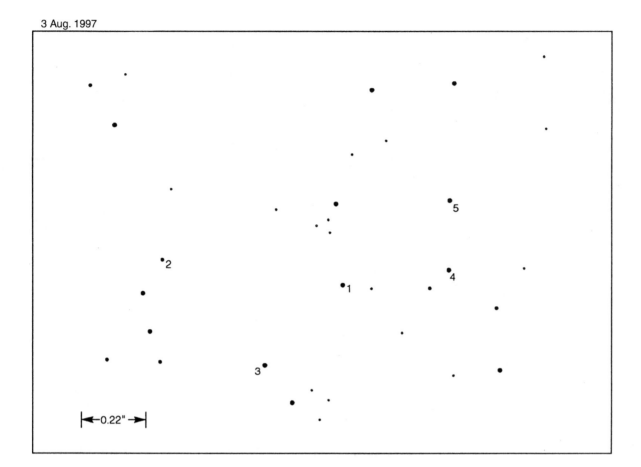

Exercise Sixteen: Measurement of Astronomical Distances

3. Under certain circumstances, it is possible to determine the individual masses of the stars in a binary system. If $m_1 = 3.1$ solar masses, what is m_2?

$m_2 =$ _____ solar masses

Binary stars are important in astronomy because they are the only systems where we can obtain stellar masses. Sometimes a star may be determined to be a binary from shifts in the wavelengths of its spectral lines, even though its individual stars cannot be resolved. Such a system is called a spectroscopic binary. Spectroscopic binaries may be detected and measured to much greater distances than visual binaries, but they yield less information.

4. If the sun had a companion star of the same mass as the sun, located 1 AU away from it, what would be their orbital period around each other? (Hint: Use our original Newtonian form of Kepler's third law.)

$P =$ _____ years

5. Why is this shorter than the period of the Earth around the sun?

IV. The Mass of the Milky Way Galaxy

Newton showed that if the matter in a system were distributed in a spherically symmetrical way, the mass *exterior* to a point at a certain distance from the sphere's center would have no net attraction on a mass at that point. This is because the gravitational attraction of the mass in one direction is exactly counteracted by the gravitation of the exterior mass in the other direction. That is, when inside a symmetrical mass distribution, only the mass within your radius contributes to your orbital motion. This principle has been extended by astronomers to the case of cylindrically symmetrical mass distributions, such as may be found in spiral galaxies.

1. The stars in a galaxy orbit the center of the galaxy under the attraction of all the mass interior to their radius. For instance, the sun orbits the center of the Milky Way galaxy with an orbital speed of about 250 km/sec. The distance to the center of the galaxy is estimated to be about 9.1 kpc (kiloparsecs). If one parsec is 2.06×10^5 AU, what is the distance (a) of the center of the galaxy in astronomical units?

   $a = $ _____ AU

2. Multiply by 2π to calculate the circumference of the sun's assumed circular orbit around the center of the Milky Way in astronomical units.

   circumference = _____ AU

3. Divide the sun's orbital speed around the galactic center (250 km/sec) by the number of kilometers in an astronomical unit (1.5×10^8) to find the sun's orbital speed in astronomical units per second.

   orbital speed = _____ AU/sec

4. Now multiply this number by the number of seconds in a year (3.15×10^7) to find the orbital speed in astronomical units per year.

   orbital speed = _____ AU/yr

5. Finally, derive the period (P) of the sun's orbit around the galactic center by dividing the circumference of its orbit by its orbital speed.

   $P = $ _____ yr

Exercise Seventeen: Kepler's Third Law and Masses in Astronomy

6. Now we can determine the mass of the Milky Way, interior to the orbit of the sun around its center. Using Kepler's third law and the a and P we have determined, find the combined mass of the sun and Milky Way:

$(m_{sun} + m_{Milky\ Way}) =$ _____ solar masses

7. How many billions of suns does this correspond to?

8. Can Kepler's third law be used with the sun's motion around the galaxy to determine the mass of the sun? Why or why not?

V. The Dark Matter Problem

1. Measurements of the orbital speeds of stars in the Milky Way and other galaxies have revealed that there must be a great deal of nonluminous mass in their outer reaches to make the rotational speeds remain high far away from the galactic centers. Assume that at a distance of 15 kpc from the center of the Milky Way, the rotational speed is still 250 km/sec. What is the distance to the center of the Milky Way in astronomical units?

$a =$ _____ AU

2. What are the circumference and orbital period P for this orbit?

circumference = _____ AU

$P =$ _____ yr

Exercise Seventeen: Kepler's Third Law and Masses in Astronomy

3. Using the new *a* and *P* you have just found, calculate the mass of the Milky Way interior to this radius.

    m = _____ solar masses

4. What percentage of the 15-kpc mass of the Milky Way lies between 9.1 and 15 kpc?

5. The great majority of the light emitted by the Milky Way comes from its central regions, not its perimeter. What can one say about the mass gravitationally revealed in the outer regions of the Milky Way?

Another instance in which the gravitational mass seems to exceed the mass of stars and gas is in clusters of galaxies. Some clusters of galaxies, if they are gravitationally bound, must contain 10 to 100 times as much nonluminous mass (dark matter) as the mass in stars and gas. The ultimate fate of the universe will depend on whether there is sufficient dark matter in space to reverse the universal expansion. More on this is contained in Exercise 28, "Radial Velocities and the Hubble Law."

Figure 17.1

Motions of the moons of Jupiter.

Source: Reprinted with permission of Sky Publishing Corp., from *Sky & Telescope* magazine.

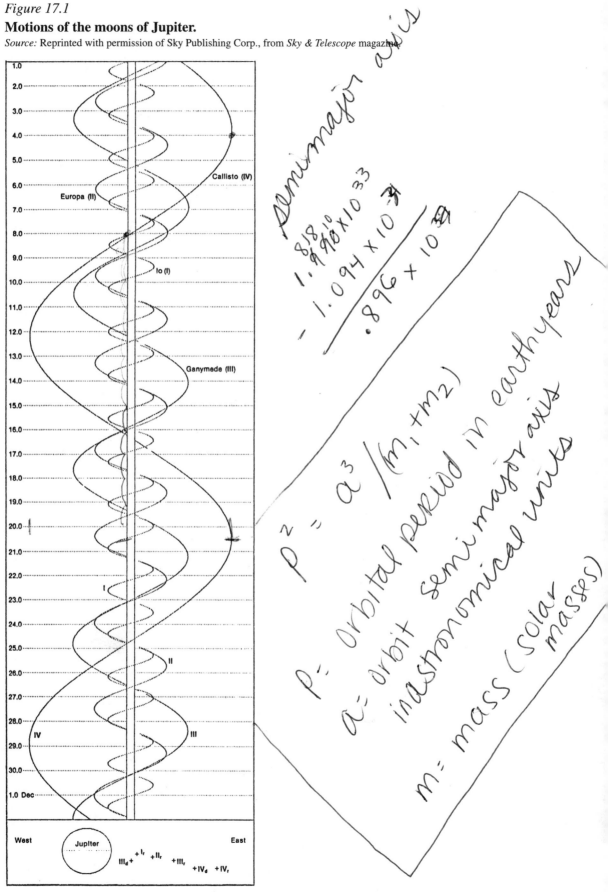

$$P^2 = a^3 / (m_1 + m_2)$$

$P =$ orbital period in earth years

$a =$ orbit semimajor axis in astronomical units

$m =$ mass (solar masses)

Name: _____

Exercise Eighteen: Outdoor

Photoelectric Photometry

I. Introduction

The quantitative measurement of the radiant energy received at the Earth from an astronomical object is known as photometry. When a photographic plate is used as the recording device, the technique is known as photographic photometry. If specially designed electronic detectors are used in place of the photographic plate, the subject is then referred to as photoelectric photometry. The heart of such a system is normally a sensitive photomultiplier that produces a small current when illuminated by light from a star or other celestial object. The current through the photomultiplier is then recorded with an external current meter or a suitable integrating device that can collect the charge that flows for an extended period of time when the signal is very weak. A typical photomultiplier is shown in Figures 18.1 and 18.2. A photon striking the photocathode C will eject an electron from the surface of the metal. This single electron is multiplied in number by the other elements of the tube, and as many as 10^6 electrons eventually cascade onto the anode A. This cascade of electrons is then the actual current measured by the external circuit. In some modem photometers, the photomultiplier is replaced by a silicon light detector.

Figure 18.1
Photomultiplier tube (cross section).

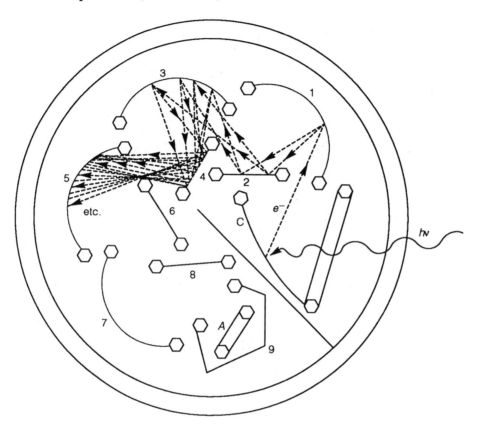

Figure 18.2
Photomultiplier tube (external view).

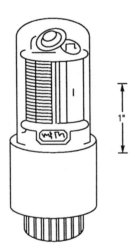

Exercise Eighteen: Photoelectric Photometry

II. Multicolor Photometry

The light reaching the detector is often limited to a particular region of the spectrum by a filter that determines the instrumental window or passband of the measurements (Figure 18.3). Several different filters for selected regions of the color spectrum can then be used to determine the brightness of an object at various wavelengths. The standard UBV system, for example, uses three filters centered on the ultraviolet (U), blue (B), and visual (V) regions of the spectrum. The effective wavelengths and passbands of the filters used in the standard UBV system are listed here.

Filter	Wavelength of Maximum Transmission	Bandwidth
U	3500 Å	700 Å
B	4350 Å	970 Å
V	5550 Å	850 Å

The relative amount of energy received in each of these instrumental windows also depends on the type of object being observed. A blue star will be brighter in U and B than it is in V because most of its energy is emitted at short wavelengths. A red star, however, will be brighter in V than in the other bands because it emits most of its radiation at long wavelengths.

Figure 18.3
Schematic diagram of a photoelectric photometer.

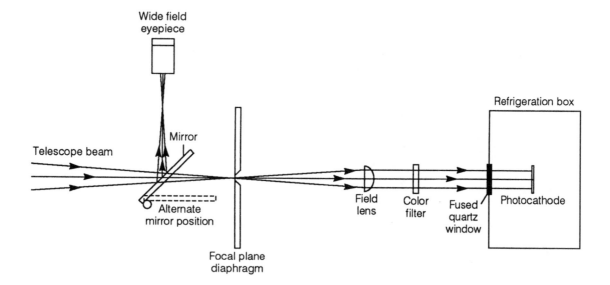

III. Instrumental Magnitudes

The measured signal from the photometer is normally converted to a magnitude scale. The instrumental magnitude of an observation is defined in terms of the logarithm of the measured signal S coming from the photomultiplier. The instrumental magnitudes of the UBV system are calculated as follows:

$$m_V = -2.5 \log S_V$$
$$m_B = -2.5 \log S_B$$
$$m_U = -2.5 \log S_U$$

Once these magnitudes are known, they must be converted to the standard UBV system. The magnitudes will then be in a form that can be interpreted by any observer familiar with the UBV system. The actual reduction of the instrumental magnitudes to the standard UBV system is usually done by observing standard stars that have known values in the UBV system. Unknown stars can be measured using the equipment once the conversion of the instrumental magnitudes to the standard UBV system is known.

IV. Observations

1. Visually observe the available standard stars in Table 18.1 and in Figures 18.4 and 18.5. List the stars below in order of decreasing brightness to the unaided eye.

    _____

    _____

    _____

    _____

    _____

    _____

    _____

    _____

    _____

    _____

    _____

Exercise Eighteen: Photoelectric Photometry

Table 18.1

Standard Stars

Star	Standard B	Standard V
Standard Stars in Auriga		
α Aurigae	0.89	0.80
γ Aurigae (β Tauri)	1.52	1.65
η Aurigae	2.99	3.17
ι Aurigae	4.19	2.66
κ Aurigae	5.36	4.34
Standard Stars in the Early Fall		
α Lyrae	0.04	0.04
θ Lyrae	5.61	4.35
α Aquilae	0.99	0.77
γ Aquilae	4.14	2.62
α Cygni	1.35	1.26

2. Measure the standard stars in B and V using a photoelectric photometer. Record the measures in the chart provided.

	Star Name	*Signal (B)*	*Signal (V)*
a.	_____	_____	_____
b.	_____	_____	_____
c.	_____	_____	_____
d.	_____	_____	_____
e.	_____	_____	_____

Figure 18.4
Finding chart 1 for standard stars.

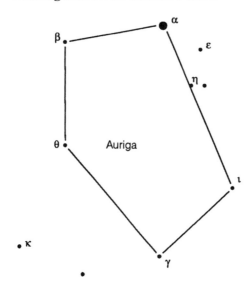

Figure 18.5
Finding chart 2 for standard stars.

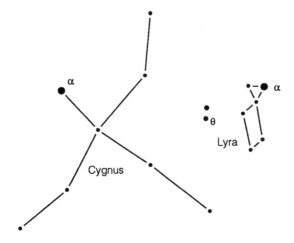

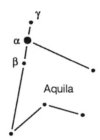

3. Measure several unknown stars in B and V.

	Star Name	Signal (B)	Signal (V)
a.	_____	_____	_____
b.	_____	_____	_____
c.	_____	_____	_____

4. Calculate the instrumental magnitudes for the standard star observations by using the equations in Part III.

	Star Name	Signal (B)	m_B	Signal (V)	m_V
a.	_____	_____	_____	_____	_____
b.	_____	_____	_____	_____	_____
c.	_____	_____	_____	_____	_____
d.	_____	_____	_____	_____	_____
e.	_____	_____	_____	_____	_____

5. Plot the instrumental magnitudes m_V for the standard stars as a function of their known V magnitudes in the UBV system on the graph paper provided.

6. Plot the instrumental magnitudes m_B for the standard stars as a function of their known B magnitudes in the UBV system on the second page of graph paper.

Exercise Eighteen: Photoelectric Photometry

7. Calculate the instrumental magnitudes m_V and m_B for the unknown stars that were also observed with the photometer.

Star Name	Signal (B)	m_B	Signal (V)	m_V
a.				
b.				
c.				

8. Using the graphs as a calibration, determine the B and V magnitudes of the unknown stars in the UBV system.

Star Name	B	V
a.		
b.		
c.		

9. Calculate the difference B – V for each of the unknown stars. This difference is known as the color index of the star.

Star Name	B – V
a.	
b.	
c.	

10. Do you expect the color index B – V to increase or decrease as the temperature of a star is raised?

Exercise Eighteen: Photoelectric Photometry

Name: *Michael Leone*

Exercise Nineteen: Indoor/Outdoor

Spectroscopy in Astronomy

In this lab you will observe examples pertaining to Kirchhoff's three laws of spectral analysis. You will use a simple prism spectroscope to observe continuous, emission, and absorption spectra.

I. The Spectroscope

A spectroscope is a device used by astronomers (and others) to separate light into its various color components. Basically, there are two types of spectroscopes; one uses a prism, usually of glass, the other uses a diffraction grating which is made of a plate of glass with very fine and accurately spaced scratches on one face. The grating or prism in a spectroscope is called the dispersing element. Your instructor will show you examples of diffraction gratings and prisms.

1. A spectroscope will be set up for your inspection. Make a schematic drawing of it, noting the position of the source to be examined, the entrance slit, the prism or diffraction grating, and the imaging eyepiece. Indicate on your drawing the path of a ray of white light from the source through the spectroscope. What do you think are the roles of the entrance slit and the imaging eyepiece?

If the spectroscope examined used a grating (prism) as a dispersing element, it would take the place of the prism (grating) in your diagram.

II. Kinds of Spectra

Kirchhoff's laws describe the conditions necessary for the observation of the three different types of spectra; continuous, emission, and absorption.

- *Continuous spectra:* A luminous solid, liquid, or very dense gas will emit light at all wavelengths, producing a continuous spectrum.

- *Emission spectra:* A rarefied (not dense) luminous gas will emit light at only certain wavelengths. Such spectra appear as bright lines superposed on a black background or on a faint continuous spectrum.

- *Absorption spectra:* If white light from a continuous spectral source is passed through a rarefied cool gas, the gas will subtract certain wavelengths from the continuous spectrum. Such spectra appear as dark lines superposed on a continuous spectrum. These dark lines appear at the same wavelengths as the emission lines would if the same gas were luminous. The wavelengths of these lines correspond to atomic transitions within the atoms of gas.

III. Physical Observations

Continuous Spectra

1. Observe a light bulb with a hand-held spectroscope. Open the slit up wide, and then focus the eyepiece so the edges of the spectrum are sharp. Narrow the slit until the spectrum is easily visible, but not too bright. The spectroscope is now focused. Can you explain why a light bulb produces a continuous spectrum?

Exercise Nineteen: Spectroscopy in Astronomy

Name: _____

Exercise Twenty: Indoor

Spectral Classification

I. Astronomical Use

Astronomers use spectral classification to find out many things about stars and the like. Temperature, composition, and surface gravity are among the characteristics of stars that astronomers can investigate by stellar spectral classification.

Your instructor will explain the use of two instruments astronomers use to help them interpret stellar spectra:

- *Spectral comparator:* a device that allows close comparison of two different spectra.
- *Microdensitometer:* an instrument that allows the astronomer to examine the spectrum of one object in detail.

II. Visual Classification

In a previous lab, you were asked to draw the spectra of four bright stars.

1. Using the fact that the colors of the spectrum correspond approximately to the following wavelengths, draw the spectra you observed again, with the lines at approximately the correct wavelengths.

Violet	4000 to 4350 Å
Blue	4350 to 4700 Å
Green	4700 to 5700 Å
Yellow	5700 to 5900 Å
Orange	5900 to 6100 Å
Red	6100 to 6900 Å

β Orionis (Rigel)
    Alternate: α Lyrae (Vega)

4000 Å   5000 Å   6000 Å   7000 Å

α Canis Minoris (Procyon)
    Alternate: α Aquilae (Altair)

4000 Å   5000 Å   6000 Å   7000 Å

α Aurigae (Capella)
    Alternate: β Geminorum (Pollux)

4000 Å   5000 Å   6000 Å   7000 Å

α Orionis (Betelgeuse)
    Alternate: α Scorpii (Antares)

4000 Å   5000 Å   6000 Å   7000 Å

Exercise Twenty: Spectral Classification

2. Several pages from the *Atlas of Stellar Spectra* are included at the end of this lab for you to examine. Notice that the wavelengths covered in the atlas extend only from about 3800 Å to about 4900 Å (ultraviolet to green). Using these pages and this limited region of the spectra you have drawn, try to classify the spectra of the four stars. (A rough classification, within one letter of classification, will be more than sufficient.) Disregard luminosity classifications. Circle the correct star names and give your classifications here:

Star	*Spectral Class*
Rigel or Vega	_____
Procyon or Altair	_____
Capella or Pollux	_____
Betelgeuse or Antares	_____

3. Which was most difficult to classify?

4. Which was easiest?

5. Do you think that visual stellar classification is very precise? Why or why not?

6. Do you think most people in your class would come up with similar classifications for the same star?

Exercise Twenty: Spectral Classification

III. Classification of Spectrograrns

1. Figure 20.1 shows several stellar spectrograms (photographically recorded spectra) to classify. Using the pages from the *Atlas of Stellar Spectra* at the end of this lab, try to match each spectrum as closely as possible with one in the atlas, and take the classification of the matching spectrum in the atlas as the spectral classification of that star. Warning— the scale of Figure 20.1 is not identical to that of the Atlas pages. Be very careful! Many spectra differ only in the *relative* intensities of a few lines. Again, disregard luminosity classifications. Write your classifications here:

Star Name	Classification	Star Name	Classification
_____	_____	_____	_____
_____	_____	_____	_____
_____	_____	_____	_____
_____	_____	_____	_____
_____	_____		

2. Now that you have classified a few spectra, a brief explanation is in order. Basically, you classified your stars by temperature. The possible spectral classes for normal stars go from O1 (very hot) to M9 (rather cold), with the order of decreasing temperature being O B A F G K M (mnemonic: Oh, be a fine girl, kiss me!). The classifications of a few stars of interest are β Orionis: B8; α Canis Minoris: F5; α Aurigae: G8; α Orionis: M2; and the sun: G2. Which of these other stars is closest to the sun in temperature?

Stars may also be classified by luminosity class, but we will leave that for the experts.

Figure 20.1

Stellar spectrograms to classify.
Source: W. W. Morgan et al., *An Atlas of Stellar Spectra*, 1943. Reproduced by permission of The University of Chicago Press.

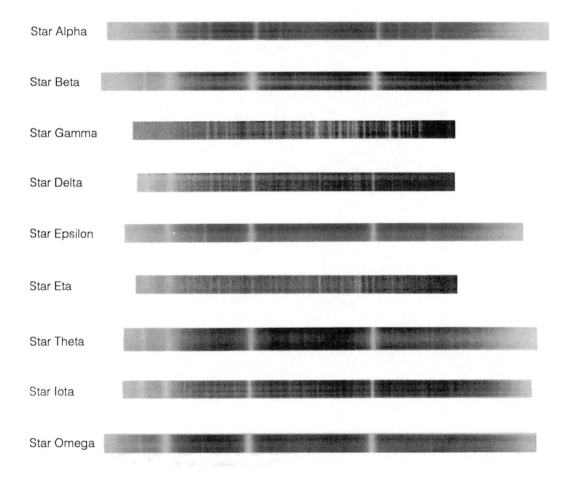

Star Alpha

Star Beta

Star Gamma

Star Delta

Star Epsilon

Star Eta

Star Theta

Star Iota

Star Omega

IV. Examination of Other Spectra

Look at the spectra of a Seyfert galaxy (Figure 20.2), of the nucleus of a normal galaxy (Figure 20.3), and of a long-period variable star (Figure 20.4). Also, examine the objective prism spectra (Figure 20.5).

1. The long streak with blobs in Figure 20.2 is the spectrum of a Seyfert galaxy. What is peculiar about the spectrum of the Seyfert galaxy?

2. What kind of material produces this kind of spectrum, according to Kirchhoff's laws?

3. Note the H and K lines of calcium in the spectrum of the nucleus of the normal galaxy. What kind of stars have strong H and K lines? (A broad category is sufficient.)

4. What can you infer about the stars that give off most of the light from the nucleus of a typical galaxy?

5. Suppose that the H and K lines of a galaxy's spectrum appear tilted. Can you think of why this might be? (Hint: Consider the Doppler effect.)

6. What unusual features do you notice in the spectrum of the long-period variable (that is, features not usually found in the spectra of normal stars)?

Exercise Twenty: Spectral Classification

7. Try to classify the spectrum of the long-period variable. Is it a hot or cool star?

8. What conclusions can you draw about this type of variable star?

9. What is the obvious advantage of objective prism plates over single-slit spectral plates?

10. What disadvantages would they have?

Figure 20.2

Image and spectrum of a Seyfert galaxy (negative prints).

Source: C. K. Seyfert, *Astrophysical Journal*, 1943, vol. 97, p. 28. Published by The University of Chicago Press.

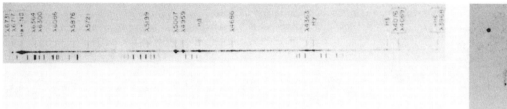 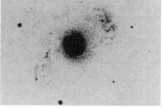

Figure 20.3

Image and spectrum of a normal galaxy (positive prints).

Source: Palomar Observatory Photographs.

 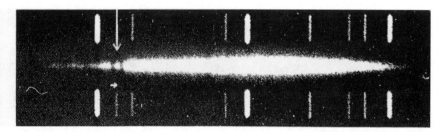

Figure 20.4

Spectrum of a long-period variable (negative print).

Source: W. W. Morgan et al., *An Atlas of Stellar Spectra*, 1943. Reproduced by permission of The University of Chicago Press.

Figure 20.5

Objective prism spectra taken with the Burrell 24-inch Schmidt. Emission line star is a Wolf-Rayet star.

Source: Courtesy of Warner and Swasey Observatory, Case Western Reserve University.

Exercise Twenty: Spectral Classification

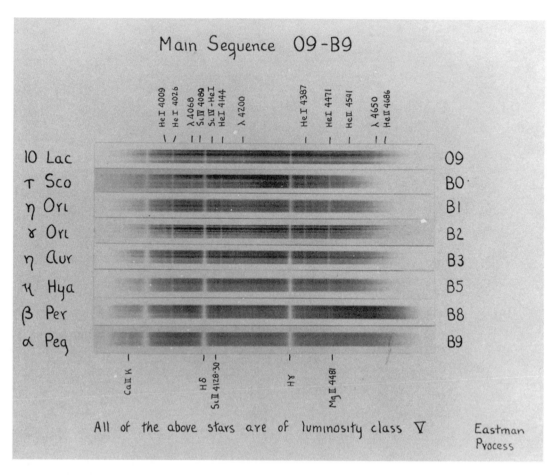

Source: W. W. Morgan et al., *An Atlas of Stellar Spectra*, 1943. Reproduced by permission of The University of Chicago Press.

Exercise Twenty: Spectral Classification

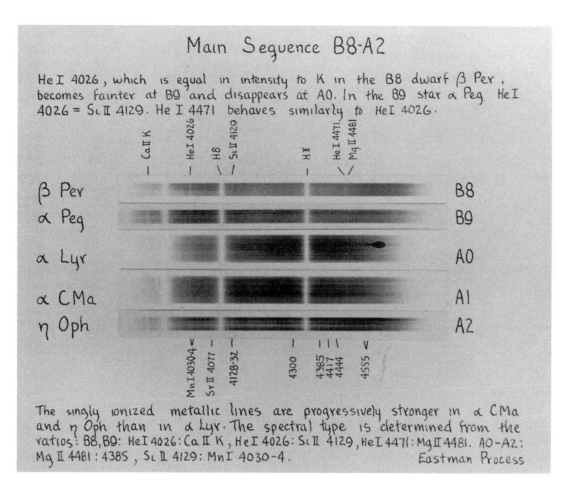

Source: W. W. Morgan et al., *An Atlas of Stellar Spectra*, 1943. Reproduced by permission
of The University of Chicago Press.

Exercise Twenty: Spectral Classification

Source: W. W. Morgan et al., *An Atlas of Stellar Spectra*, 1943. Reproduced by permission of The University of Chicago Press.

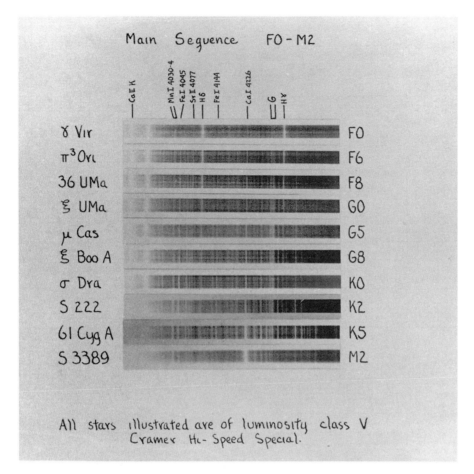

Source: W. W. Morgan et al., *An Atlas of Stellar Spectra*, 1943. Reproduced by permission of The University of Chicago Press.

Exercise Twenty: Spectral Classification

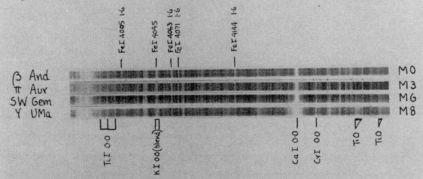

The M Sequence Is A Temperature Sequence

The spectral types of the four stars were determined from the TiO bands in the blue-green region (not shown in the illustration). The arrangement in order of increasing band absorption has the following characteristics: the strong FeI lines having excitation potentials of around 1.6 volts grow systematically weaker; ultimate lines of TiI, CaI and CrI grow systematically stronger;

the KI pair situated on each side of FeI 4045, and blended with it on the spectra shown, become systematically stronger. This is shown by the change in the width of the line from M0 to M8. These changes indicate that stars having the strongest bands have also the lowest excitation temperatures among the M stars. They thus show that the TiO bands do not pass through a maximum of intensity. The plates were taken by Keenan with the McDonald 82-inch reflector and a spectrograph giving a dispersion of 65 A per mm at λ 4200. The progressive change in the line spectra of M giants was first described in detail by Merrill and his associates at Mount Wilson.   Agfa Super Plenachrome Press

Source: W. W. Morgan et al., *An Atlas of Stellar Spectra*, 1943. Reproduced by permission of The University of Chicago Press.

Exercise Twenty: Spectral Classification

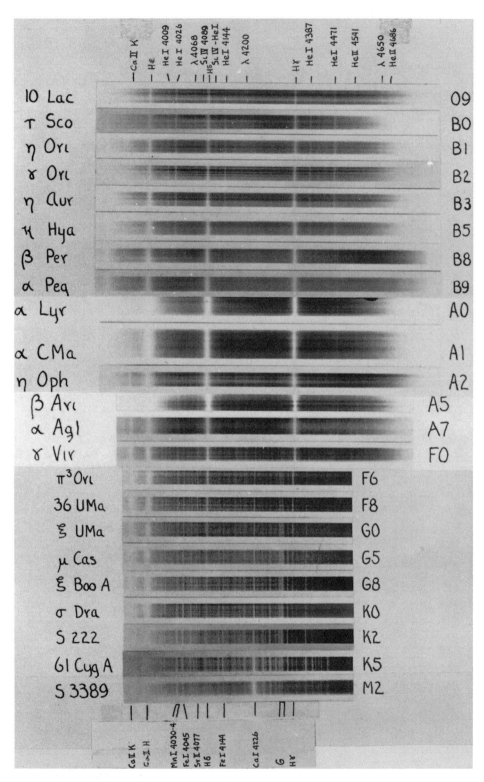

Source: W. W. Morgan et al., *An Atlas of Stellar Spectra*, 1943. Reproduced by permission of The University of Chicago Press.

Exercise Twenty: Spectral Classification

Name:

Exercise Twenty-Two: Outdoor

Telescopic Observing with Equatorial Mounting, Clock Drive, Setting Circles, and Slow Motion Controls

I. Introduction

The finest telescopes usually have an equatorial mounting, enabling them to follow the apparent motions of the stars by turning on only one axis. This makes it easy to build in a clock drive, which turns the telescope around the axis at the precise rate to cancel the motion of the Earth about its axis, and thus to track the stars in the sky. Such telescopes often have setting circles to allow them to be pointed accurately to the known positions of stellar objects, clamps to allow them to be fixed on the stellar objects found, and sometimes slow motion controls allowing them to be slightly repositioned after they have been clamped. If your institution has telescopes with equatorial mountings, this lab will instruct you in their general use.

II. Using the Telescope

If the telescope you are using is inside a dome, there will be provisions for opening the dome and turning it to follow the telescope. Your instructor will show you how to do these things. Normally the dome is left motionless for a period of several minutes while people are observing, and then moved to keep the field of view within the slit. It is helpful if the observer can instruct an observing partner when to move the dome and give start and stop instructions if the dome controls are not handy to the observer.

Exercise Twenty-Two: Telescopic Observing

Equatorially mounted telescopes should have clamps on the right ascension and declination axes. Familiarize yourself with their locations and operation. When an object is in the field of view, the clamps should be engaged. Clock motor drives always work even when the clamps are engaged, to keep the object being observed within the field of view. Some telescopes have slow motion controls, which may be used even while the clamps are engaged. During normal observing of a single object, the clamps are tightened to keep accidental bumping or imbalance from moving the telescope away from the object of interest. When moving to another object to observe, the clamps are loosened, and the telescope may then be rapidly moved on both axes to the new position. This process is called slewing. *Never* slew the telescope with the clamps engaged. *Never* force the telescope. If a gentle motion doesn't work, one of the clamps is probably on.

III. Finding a Celestial Object with an Equatorial Telescope

Each student will be assigned an object to find with an equatorially mounted telescope. You should first look up the right ascension (RA or α) and declination (dec or δ) in *The Observer's Handbook* or the *Astronomical Almanac* for the year of observation. (*The Observer's Handbook* is published annually by the Royal Astronomical Society of Canada; the *Astronomical Almanac* is published annually by the U.S. Naval Observatory.)

The other setting circle usually measures the hour angle, whereas you have only obtained the right ascension and declination from the table. To convert right ascension into a usable hour angle you must determine the sidereal time (or star time) from a sidereal clock or computer program. Then use the following equation:

$$HA = ST - RA$$

where HA = the hour angle, which is to be set on the setting circle, ST = the sidereal time, which you obtain from a sidereal clock, and RA = the right ascension, which you found from the table.

When reading the sidereal time, it is a good idea to add on about five minutes, since it usually takes about that long to perform the calculations and point the telescope. After setting the setting circle, you may then glance at the true sidereal time again and touch up your pointing, if necessary. A positive hour angle means that the object is west of your meridian and a negative hour angle means that it is east.

Once you have set the hour angle and declination on the setting circles, the object should be in the field of view of the finder telescope. Center the object on the cross hairs of the finder using the slow motion controls (if available) if the object is not too far from the center, or using the rapid motions (remember to unclamp) otherwise. When the object is centered in the finder, it should be in the field of view of a low power eyepiece in the main telescope.

Some objects are so faint that they are not visible in the finder. In this case you must use star charts such as the *Atlas Borealis* and the *Atlas Eclipticalis* and try to pick out star patterns when looking through the finder telescope. (The *Atlas Borealis* and the *Atlas Eclipticalis* were prepared by Antonin Becvar of the Prague Observatory.) Then place the center of the cross hairs over that part of the star pattern where the object is located, in order to put the object in the field of main telescope. This technique takes quite a bit of observing experience before it can be done well.

All objects that will be assigned as part of this lab should be bright enough to be seen through the finder.

IV. Star Hopping

Another technique often used by amateurs is called "star hopping." It does not require a polar axis alignment as accurate as the other methods, and thus is more suitable for portable telescopes. For an approximate method of aligning the polar axis of a telescope, see Appendix E: Aligning a Telescope Axis.

In star hopping, for every object you are interested in finding, choose a nearby bright star. Subtract the coordinates of the bright star from the coordinates of the object you are trying to find. Then, find the bright star in your finder telescope, and center the bright star in your telescope field of view. Now, using the telescope setting circles, move the telescope so as to add the difference in coordinates you had previously determined to both its hour angle and declination. Your object should now be visible in the finder, and may be centered in the main telescope field of view.

Here is an example. Suppose you want to find the Ring Nebula in Lyra. Its coordinates are:

$\alpha = 18^h\ 53^m\ 36^s$
$\delta = 330°\ 02'\ 00''.$

The bright star you wish to use to help find the Ring Nebula is the star Alpha Lyrae. Its coordinates are:

$\alpha = 18^h\ 36^m\ 54^s$
$\delta = 38°\ 47'\ 00''.$

Subtracting these coordinates from those of the nebula, we have

$\Delta\alpha = 16^m\ 42^s$
$\Delta\delta = -5°\ 45'.$

So, after finding Alpha Lyrae in your telescope, move it forward in hour angle by $16^m\ 42^s$ and south in declination by $5°\ 45'$. The Ring Nebula should now be centered in your telescope finder.

Exercise Twenty-Two: Telescopic Observing

In order to use this method, you should have detailed star charts or a star atlas that gives coordinates of both stars and other objects.

Name: _____

Exercise Twenty-Three: Indoor

Pulsars

I. Introduction

Pulsars are rapidly varying radio stars that seem to emit radiation only in short pulses. It is believed that they are the remnants of supernova explosions and that a rotating neutron star is central to each one. In this lab you will examine the properties of pulsars, particularly those influencing our beliefs about their makeup.

Interestingly enough, there are only a few pulsars known to emit visible light pulses. The first optical pulsar, located in the Crab Nebula (M1) in Taurus, was discovered in 1969. Figure 23.1 shows the Crab Nebula with the pulsar marked.

II. Characteristics of Pulsation

The Sizes of Pulsars

1. Figure 23.2 shows a plot of observed radio radiation versus time for more than one complete period of a "typical" pulsar. How does the period of this typical pulsar compare with the period of most variable stars?

2. The short period of pulsars is one of the characteristics used by radio astronomers looking for pulsars. Going by the general rule that the shorter the period of a variable, the smaller it is, are pulsars larger or smaller than normal variable stars?

3. It is a good guideline to go by that nothing astronomical can vary significantly in a time shorter than it takes for light to travel across it. Taking 0.02 sec as a typical time for a pulsar light variation and 3×10^{10} cm/sec for the speed of light, what is the maximum size of a pulsar?

$$d < t_{var} \times c$$

4. The earth's diameter is about 1.3×10^9 cm. The moon's diameter is about 3.5×10^8 cm. Compared to the Earth and moon, what is the maximum size of a typical pulsar?

Some pulsars have a time variation of 0.002 sec. The only kind of star known that could have a major light variation in this amount of time is a neutron star, which would pack the entire mass of the sun in a region no larger than New York City.

The Pulsar "Clock"

1. Two hypotheses were initially put forward as explanations for the periodicity of pulsars. One was that pulsars might be pulsating, like normal intrinsic variable stars. The other was that they might be rotating, and a narrow beam of radiation would strike us during each rotation (the lighthouse effect). According to the first hypothesis, the period of a pulsar should remain constant (or slightly decrease) as it ages. The lighthouse effect predicts that as the rotating neutron star loses energy, it will slow down and the period should increase. It has been found that the periods of all pulsars are increasing. What do you think causes the periodicity of pulsars, pulsation or rotation?

III. Positions and Brightnesses of Pulsars

1. Figure 23.3 shows the positions of a relatively uniformly selected sample of 84 pulsars in galactic longitude (l) and latitude (b). The galactic plane lies along $b = 0$ and the sun and spiral arms of the Galaxy lie near to the galactic plane. Do most pulsars lie close to the galactic plane or far away from it?

2. We believe (from observing other galaxies) that the stars that explode to become supernovae lie in or near the spiral arms of galaxies. We also know that one pulsar lies at the center of the Crab Nebula—the remains of a supernova observed by Chinese astronomers in the year 1054 A.D. If all pulsars originate in the same way, what can we infer about their origin?

3. Assuming that all pulsars lie in the narrow disk of the Milky Way, the pulsars that appear to lie close to the galactic plane must be, *on the average,* the farthest away. At great distances, only the intrinsically brightest pulsars will be visible. Therefore, on the average, the pulsars seen near to the plane will be the intrinsically bright ones, and the pulsars away from the plane will be nearby faint ones. In Table 23.1 the pulsars in Figure 23.3 that are nearer than 6° from the plane are listed and their periods given. In Table 23.2 the pulsars apparently farther than 6° from the plane are listed and their periods given. Average the periods of the pulsars in Table 23.1 and record the average period at the bottom of the table. Do the same for the pulsars in Table 23.2. Circle your choices: The pulsars that seem close to the plane have, on the average, (longer, shorter) periods than those which seem far from the plane. Thus, the brightest pulsars have (shorter, longer) periods than average.

4. Circle your choices and fill in the blank: In conclusion, we have seen that pulsars are probably connected with (large, small), (rotating, pulsating) neutron stars. They seem to be the remnants of _____ explosions in or near the spiral arms of our galaxy. The brightest pulsars have (longer, shorter) periods than the faint ones.

Exercise Twenty-Three: Pulsars

IV. Pulsar Puzzlers (Optional)

1. The mass of the sun is 2×10^{33} gm, and its volume is 1.4×10^{33} cm^3. What is its density in gm/cm^3?

2. How does that compare to water (which has a density of 1 gm/cm^3)?

3. Suppose a pulsar has a typical mass of half the sun and a radius of 10 km (volume = 10^{19} cm^3). What is its density in gm/cm^3?

4. A thimble holds 1 cm^3 (if carefully chosen). Which would weigh more, a thimbleful of pulsar or the Queen Mary? (The Queen Mary weighs 83,000 tons; 1 ton is about 10^6 gm.) How much more?

5. The surface gravity of a pulsar is about 10^{11} times the Earth's surface gravity. The work done in lifting a weight is found by Work = Mass $\times$ Surface Gravity $\times$ Distance Lifted. Is it more work to lift 1 gm a height of 1 cm on a pulsar or to lift 1000 tons a distance of 1 m on Earth?

6. The highest mountains on a pulsar are thought to be about 1 cm high! Why do you think this is so?

Exercise Twenty-Three: Pulsars

Figure 23.1

The Crab Nebula and Crab Nebula Pulsar.

Source: Palomar Observatory Photograph.

Figure 23.2
Light curve of a "typical" pulsar of 1/2 second period.

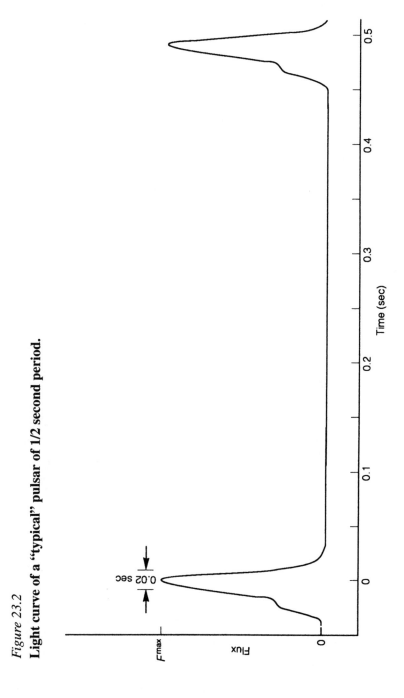

Exercise Twenty-Three: Pulsars

Figure 23.3
A uniform sample of 84 pulsars.

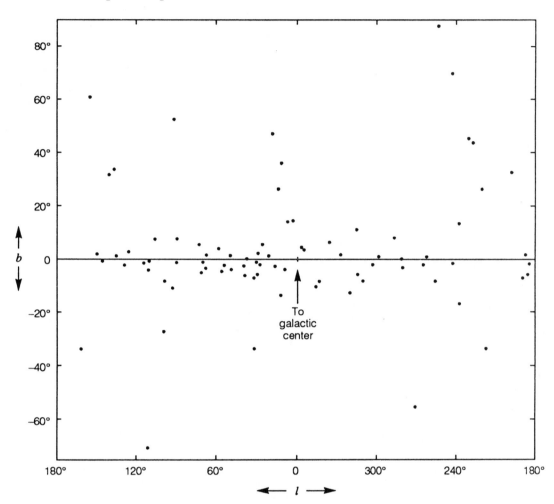

Exercise Twenty-Three: Pulsars

Table 23.1

Pulsars Appearing Nearer Than 6° from the Galactic Plane

Pulsar	b (°)	Period (sec)	
0105+65	3.3	1.28	
0138+59	−2.3	1.22	
0153+61	0.2	2.35	
0329+54	−1.2	0.71	
0355+54	0.9	0.16	
0531+21	−5.8	0.03	
0540+23	−3.3	0.25	
0611+22	2.4	0.33	
0740−28	−2.4	0.17	
0833−45	−2.8	0.09	
0835−41	−0.3	0.77	
0940−56	−2.5	0.66	
0959−54	0.3	1.44	
1154−62	−0.2	0.40	
1240−64	−1.6	0.39	
1449−65	−5.3	0.18	
1530−53	1.9	1.37	
1717−29	4.3	0.62	
1718−32	2.5	0.48	
1749−28	−1.0	0.56	
1818−04	4.7	0.60	
1819−22	−4.3	1.87	
1822−09	1.3	0.77	
1826−17	−3.3	0.31	
1831−03	2.3	0.69	
1845−01	0.2	0.66	
1845−04	−1.0	0.60	
1846−06	−2.4	1.45	
1858+03	−0.6	0.66	
1900−06	−5.6	0.43	
1907+02	−2.7	0.49	
1915+13	0.6	0.19	
1919+21	3.5	1.34	
1929+10	−3.9	0.23	
1933+16	−2.1	0.36	
1944+17	−3.5	0.44	
1946+35	5.0	0.72	
1953+29	0.7	0.43	
2002+30	−0.7	2.11	
2016+28	−4.0	0.56	
2020+29	−4.7	0.34	
2111+46	−1.3	1.01	
2256+58	−0.7	0.37	
2305+55	−4.2	0.48	Average Period
2319+60	−0.6	2.26	

Exercise Twenty-Three: Pulsars

Table 23.2

Pulsars Appearing 6° or More from the Galactic Plane

Pulsar	b (°)	Period (sec)
0031–07	−69.8	0.94
0254–54	−54.9	0.45
0301+19	−33.2	1.39
0450–18	−34.1	0.55
0525+21	−6.9	3.75
0628–28	−16.8	1.24
0736–40	−9.2	0.37
0809+74	31.6	1.29
0818–13	12.6	1.24
0823+26	31.7	0.53
0834+06	26.3	1.27
0904+77	33.7	1.58
0943+10	43.2	1.10
0950+08	43.7	0.25
1055–51	7.0	0.20
1112+50	60.7	1.66
1133+16	69.2	1.19
1237+25	86.5	1.38
1359–50	11.0	0.69
1426–66	−6.3	0.79
1451–68	−8.6	0.26
1508+55	52.3	0.74
1541+09	45.8	0.75
1556–44	6.4	0.26
1604–00	35.5	0.42
1642–03	26.1	0.39
1700–18	14.0	0.80
1706–16	13.7	0.65
1727–47	−7.6	0.83
1747–46	−10.2	0.74
1857–26	−13.5	0.61
1911–04	−7.1	0.83
1917+00	−6.1	1.27
2021+51	8.4	0.53
2045–16	−33.1	1.96
2148+63	7.4	0.38
2154+40	−11.4	1.53
2217+47	−7.6	0.54
2303+30	−26.7	1.58

Average Period _____

Exercise Twenty-Three: Pulsars

Name: _____

Exercise Twenty-Four: Indoor

Galactic Spiral Structure

I. Introduction

Spiral galaxies were first observed visually with large telescopes during the second half of the nineteenth century. The introduction of photography showed how common these objects are. Their true distances and nature were not revealed until 1926, when it was shown that they are stellar systems external to our own.

It is easy enough to trace the spiral structure of an external galaxy. All you have to do is examine photographs of it. The task of delineating the structure of our own Galaxy is much more difficult. As you might imagine, it is far easier to map a city from an airplane than from one of its street corners. Thus the oldest picture of galactic spiral structure dates back only to 1951.

Examine pictures of other spiral galaxies in Exercise 27, "Classification of Galaxies." The spiral structure is revealed by certain types of objects found mainly in the spiral arms. These objects are O and B type stars and the clusters and associations they are found in, clouds of ionized hydrogen (HII regions), very bright Cepheid variables, and other supergiant stars. It is now thought that stars and clusters are formed in the spiral arms. Then they drift away, typically at a speed of 10 pc/million yr. Only very young objects, such as those just listed, will still be found near their point of origin.

The problem with such work, accomplished optically, is that the dust in our Galaxy makes it impossible to observe objects in the galactic plane much more than 5000 pc away. But spiral galaxies are 30,000 to 40,000 pc in diameter, so we cannot see very far in our own Galaxy at all.

Radio astronomy has come to our aid. Radio waves can penetrate the dust and see objects (such as clouds of hydrogen gas) as far away as 20,000 pc. The big problem in radio work is that the distances of the gas clouds we observe are difficult to determine, so it is hard to see how they fit into a spiral pattern.

In this lab exercise you will use the distances and positions of optical objects to trace the spiral structure of the Galaxy. You will then be able to compare it with the radio results and get an idea of the problems involved.

II. Optical Spiral Structure

The accompanying lists, Tables 24.1 through 24.3, give the distances and galactic longitudes of objects useful in spiral tracing. Using polar coordinates graph paper, plot a group of these objects specified by your instructor. (Remember that 0° of galactic longitude is in the direction of the Galactic center and that longitude increases counterclockwise.) Because the galactic latitudes of these objects are nearly zero, the latitudes have been neglected. The scale to use is 1 inch = 1 kpc (kiloparsec).

After you have plotted your group of objects, the instructor will combine all the plots into a master diagram of all types of spiral tracers. On this plot you should be able to see how the objects tend to clump together along certain favored curves. These mark our Galaxy's optical spiral structure.

1. Looking at the plot of all spiral tracers, draw light pencil lines on your own plot to mark what you think are the important spiral features. Remember that the galactic center is about 5 kpc below the bottom of your plot, and you are located at the center of your plot.

2. There are occasional gaps in the arms. Do you think these are genuine gaps or just regions of observational deficiency? Before answering this, be sure to look at some of the spiral galaxies in Exercise 27.

Exercise Twenty-Four: Galactic Spiral Structure

Table 24.1

Supergiants

Star	l (°)	r (kpc)	Star	l (°)	r (kpc)	Star	l (°)	r (kpc)
HD170938	16.8	1.79	HD 6182	124.4	2.36	CD − 44°3129	264.7	3.12
HD171094	18.3	2.37	BD + 62°246	126.7	3.48	HD80558	273.1	1.28
HD173783	25.1	4.04	BD + 62°297	128.7	2.87	HD91619	285.6	2.26
UW Aql	34.0	1.96	HD13476	133.5	2.22	HD94493	289.0	3.06
BD + 23°3745	59.5	2.19	HD14242	134.2	2.29	HD98733	291.9	3.18
BD + 31°3921	69.4	2.96	HD14134	134.6	2.00	HD101332	294.8	3.02
BD + 33°4077	74.2	1.76	BD + 56°595	135.1	2.50	HD106343	298.9	2.75
HD192660	77.4	1.61	HD15316	135.8	2.49	HD111973	303.2	2.15
HD228882	78.1	1.91	YZ Per	137.1	2.31	HD115363	305.9	5.01
HD195592	82.4	0.90	BD + 57°647	138.6	2.31	HD122879	312.3	1.97
HD205196	98.6	1.37	HD21291	141.5	1.03	HD142565	328.0	3.45
μ Cep	100.6	0.58	BD + 52°729	148.0	3.48	HD329905	330.4	5.57
HD239895	103.0	3.28	χ Aur	175.8	0.91	HD328209	338.5	3.28
HD213470	104.7	2.49	HD40589	182.9	2.30	HD152147	343.2	2.05
HD218915	108.1	3.05	α Ori	199.8	0.14	HD161653	352.4	3.09
TZ Cas	115.9	2.78	τ Ori	214.5	0.44	HD163065	359.7	4.09
HD225146	117.2	3.09	VY CMa	239.4	2.98	HD100198	293.4	2.46

Table 24.2

Young Open Clusters

Cluster	l (°)	r (kpc)	Cluster	l (°)	r (kpc)	Cluster	l (°)	r (kpc)
K 14	121	2.40	NGC 2244	206	1.60	I Ara	339	1.40
NGC 457	127	2.75	NGC 2264	203	0.70	NGC 6231	344	1.75
M 103	128	2.45	NGC 2353	225	1.30	NGC 6322	345	1.10
Tr I	128	2.45	NGC 2362	238	1.55	NGC 6530	006	1.55
NGC 637	129	2.10	NGC 2439	246	1.55	M 21	008	1.35
NGC 654	129	2.55	NGC 2467	243	2.50	M 16	017	1.70
NGC 659	129	2.10	II Pup	244	2.85	NGC 6823	059	2.40
NGC 663	130	2.15	NGC 2546	255	0.85	NGC 6871	073	1.60
h Per	135	2.15	IC 2581	285	2.75	IC 4996	075	1.80
χ Per	135	2.50	IC 2602	290	0.15	NGC 6910	079	1.60
IC 1805	135	2.40	Tr 14	287	1.65	M 29	077	1.25
NGC 957	136	2.10	Tr 15	287	1.70	IC 1396	099	0.70
IC 1848	137	2.30	Tr 16	288	2.96	NGC 7128	097	3.10
α Per	147	0.17	NGC 3766	294	1.80	IC 5146	094	0.95
II Per	160	0.40	IC 2944	295	1.95	NGC 7160	104	0.70
NGC 1444	148	0.93	NGC 4103	298	1.90	I Lac	097	0.55
NGC 1502	144	0.85	Pi 20	321	1.45	NGC 7380	107	2.80
I Ori	207	0.40	II Sco	352	0.15	III Cep	110	0.70
NGC 2129	187	1.85	NGC 6193	337	1.30	NGC 7510	111	2.90
NGC 2169	196	0.95	NGC 6204	339	1.30	V Cas	116	2.00
						NGC 7788	116	2.40

Table 24.3

OB Associations

Association	l (°)	r (kpc)	Association	l (°)	r (kpc)	Association	l (°)	r (kpc)
Sgr OB1	9	1.56	Cyg OB4	82.5	1.00	Per OB3	147	0.17
Sgr OB7	10.6	1.86	Cyg OB7	90	0.74	Cam OB3	147.0	3.5
Sgr OB4	12.0	2.13	Cep OB2	102.5	0.70	Aur OB1	173.1	1.34
Ser OB1	16.5	1.70	Cep OB1	103	3.60	Gam OB1	189.1	1.50
Sct OB3	17.4	1.60	Cep OB5	108.4	2.09	Mon OB1	203	0.72
Ser OB2	18.6	2.00	Cas OB2	112.0	2.68	Mon OB2	207	1.40
Vul OB1	60.3	2.05	Cas OB4	120.3	2.65	CMa OB1	224	1.32
Vul OB4	60.5	1.02	Cas OB14	120.4	1.18	Vel OB1	265	1.43
Cyg OB3	72.6	2.30	Cas OB7	123.4	2.34	Car OB1	287.5	2.60
Cyg OB1	75.5	1.70	Cas OB1	124.0	2.63	Cen OB1	304.5	1.50
Cyg OB8	77.8	2.19	Cas OB8	129.5	2.9	Sco OB1	343.3	1.40
Cyg OB9	78	1.17	Cas OB6	135.9	2.42	Sgr OB5	0.2	2.60
Cyg OB2	80.1	1.50	Cam OB1	142.5	0.90			

3. Using a dashed pencil line, draw in some possible alternative interpretations of how the different clumps connect with each other. Examine the gap around galactic longitude 320° especially.

4. On the scale of your diagram, how many inches across is the Galaxy, if it is 30 kpc in diameter? How much of its area have we really mapped?

III. Optical–Radio Comparison

1. Figure 24.1, illustrating the radio structure of our Galaxy, has a box in the upper half representing the area available on your polar coordinate paper. With this indication of scale, plot your optical spiral structure lines on this drawing. Just plot the lines, not individual objects. This will give you a better indication of how far out the Galaxy has been optically mapped.

2. Does your optical spiral structure match the radio structure?

3. What are the discrepancies, if any?

4. Make a separate drawing in the space provided of the major features of our Galaxy as it might be seen edge on.

5. Why can't we use old objects such as globular clusters to trace spiral structure?

6. Our Galaxy is very thin, almost all of its stars lying very close to its plane. Why can we observe some stars in a direction perpendicular to the plane? (Hint: consider the thickness of the plane.) If we see equal numbers of stars above and below the plane, what can we say about the sun's position in the plane?

7. (Optional) An RR Lyrae variable star has an absolute magnitude of $M_v = +0.6$. It is observed through 4.3 magnitudes of absorption and is apparently located very near the galactic center. If its apparent magnitude is $m_v = +19.5$, how far away is the galactic center? Use the distance relation

$$m_v = M_v + 5 \log D - 5 + A_v$$

8. What happens to this estimate (does it get larger or smaller) if $M_v = +1.0$?

Exercise Twenty-Four: Galactic Spiral Structure

Figure 24.1

Radio structure of our Galaxy. The position of our sun is marked by the small circle. The rectangle represents the limits of the graph paper used in plotting the optical structure.

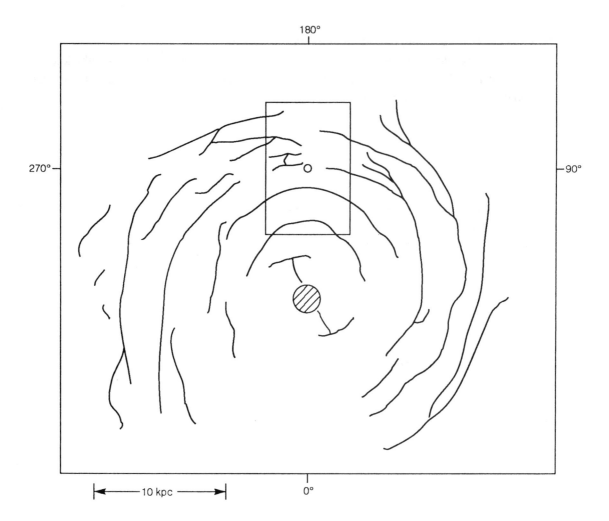

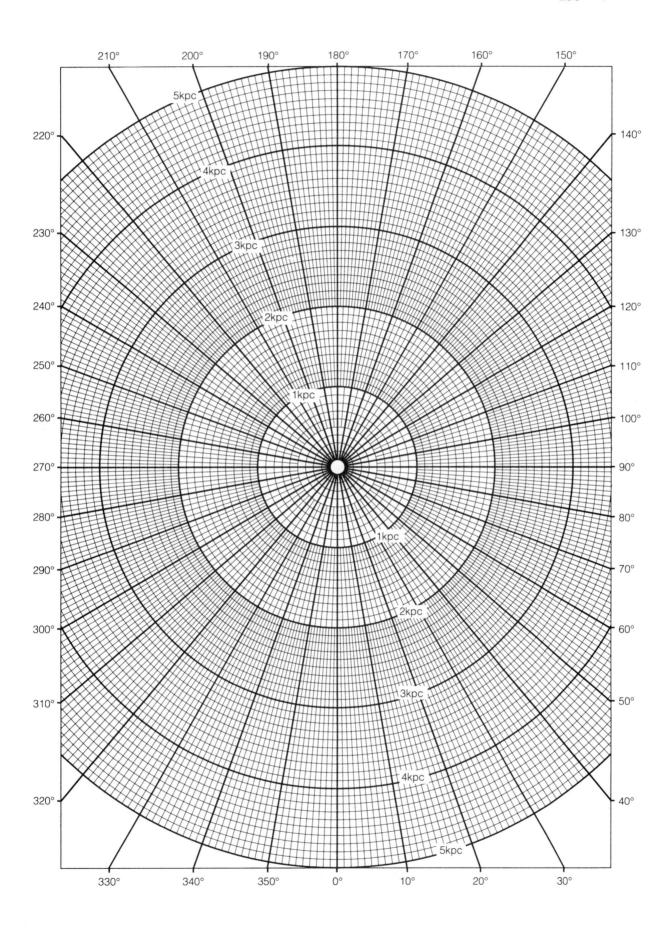

Name: _____

Exercise Twenty-Five: Indoor

Astronomical Image Processing

I. Introduction to Astronomical Images

Most astronomical images are now obtained by using CCD (charge coupled device) cameras. CCD camera chips are much more sensitive to light than photographic emulsions, typically recording up to one half of all photons striking them, rather than the fraction of one percent recorded by regular photography. Thus, CCD images make much more efficient use of the light at the telescope focus than do photographs. A CCD chip is a rectangular array of small light-sensitive areas called pixels. When a photon strikes a pixel, an electron is generated and stored in the pixel. After the exposure is completed, all of the pixels are read out one at a time, and the number of electrons in each is recorded in an array of numbers, and stored in a computer file, called an image file. Now, each number in the array is called a pixel value. Usually, there are a limited number of pixels in the CCD image, such as 192 rows by 165 columns. When the image is displayed, the computer screen illumination is varied to correspond to the pixel values, and the rows and columns are displayed in the same relative locations as they were recorded in the CCD camera to make up the image.

CCD cameras offer many advantages over conventional photography. In addition to being much faster (i.e., more sensitive), CCD images usually are very linear (this means the pixel values are tied very closely to the number of photons incident), whereas photography is plagued with sensitivity that is variable depending on the exposure time, the total light intensity, etc. CCD images require no chemical processing. Because they can be displayed immediately upon their acquisition, only the best exposures need be permanently stored. Storage requires only limited space on computer hard drives or floppy disks. An infinite number of copies of the images could in principle be made, each identical to the original.

However, CCD images also have some disadvantages. Often, the CCD chip is very small, and so the focus of the camera is extremely critical. Telescope pointing and tracking are also critical, with errors in either leading to off-center and/or blurred images. The individual pixels in a CCD camera may not have identical light sensitivities, leading to a speckled appearance even for uniform illumination. In addition, electrical noise can lead to random brightness variations between pixels. Even in the absence of light, electrical noise leads to electrons occupying pixels all over the image. Usually, CCDs are cooled to reduce electrical noise, but temperature differences across the array then may lead to apparent brightness variations in the image. Finally, cosmic rays can strike the camera pixels, leading to a large number of electrons being generated in some individual pixels, independent of illumination.

Because CCD images are stored as arrays of numbers, mathematical techniques can be used to improve their appearance and to help display the information they contain. In this exercise, you will learn to use computer software to implement some of these techniques, and learn to optimize the appearance of CCD astronomical images.

II. Computer Software Techniques

Many software programs have been developed to process astronomical (and non-scientific) images. In this exercise, we will use terms that may apply to any software package. One of our favorite shareware programs is called Graphic Workshop, by Alchemy Mindworks, Inc., of Ontario, Canada. After downloading and trying it out, a registered version may be purchased from the publisher. Another excellent shareware program is Paint Shop Pro, by JASC, Inc., of Minnetonka, Minnesota. Many others will work as well. An "undo" capability comes in very handy sometimes in image processing. Some CCD makers include or distribute software to use with their cameras. It may be quite sophisticated, and may include all the processing functions used in this exercise. Other software packages offer similar image processing techniques, sometimes with slightly different terms for the same techniques. If you are using one of them, substitute its different terms for those used in this exercise.

Subtracting the Dark Image and "Flat-Fielding"

The effects of differences in background noise between pixels and across the CCD chip and many other effects can be corrected for by subtracting an image called a dark image from the image containing the desired object. The dark image is obtained before or after the telescopic image, using the same camera and the same exposure time as the telescope image, but without opening the camera shutter. The dark image contains only "noise," which must also be present in the telescopic image, and which can be subtracted using image processing. Most CCD camera software includes a dark image subtraction feature. Another useful feature included in most CCD camera software is called "flat-fielding." Because the sensitivity of a CCD camera can vary across the field of view, an image of a uniformly bright field of view can be used to "flatten" the response of the image pixels. See your CCD camera manual for details.

Figure 25.1 below shows a sum of four images taken with the 10-cm finder telescope of the Baldwin-Wallace College Burrell Memorial Observatory 33-cm refractor on a historic occasion. It shows a supernova near the central bulge of the Whirlpool galaxy M51, as seen on April 16, 1994. The supernova is the smaller of the two central bright spots, and the galaxy's central bulge is the brighter of the two spots. As you can see, the images were exposed to highlight supernova 1994I and it takes some concentration to see the spiral structure of M51. In this laboratory exercise, we will see how to bring out and accentuate the spiral structure of M51 from pixels that were underexposed in this summed image, which may be obtained on-line from Brooks/Cole with the title 94isum.tif.

Figure 25.1

Stretching the Image Contrast

While Figure 25.1 shows the galaxy M51, the image is still not very distinct, for lack of contrast between the dark and light parts of the image. One of the important steps in image processing is to improve the contrast by making the darkest parts of the image appear very dark, and the lightest parts appear very light. This is most easily done by viewing a histogram (a bar chart) of the numbers saved in all the pixels of the image. The minimum saved pixel value in the histogram can then be set to zero, the maximum saved value set to the maximum allowed value, and the intervening values are interpolated to values in

Exercise Twenty-Five: Astronomical Image Processing

between these. The histogram for Figure 25.1 is shown below as Figure 25.2. At the right are the very few pixels which are very bright, the ones that contain the image data for the central bulge and supernova.

Figure 25.2

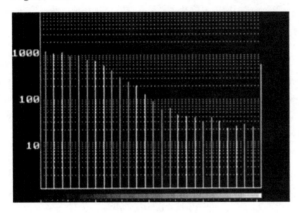

While the pixels at the right contain the supernova, most of the pixels at the very left of the histogram contain noise and the faint spiral structure of M51.

1. Using your image processing software on this image, set the black pixel value to be about 100 counts, and the white pixel value to about 600. Now, your histogram should look like Figure 25.3 below. If your software doesn't show histograms, adjust your contrast and brightness to produce an image like Figure 25.4.

Figure 25.3

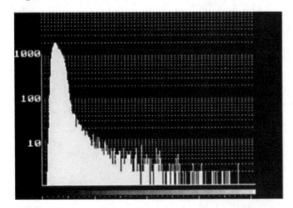

After this "stretching" operation (sometimes called adjusting contrast and brightness), the result is Figure 25.4.

Figure 25.4

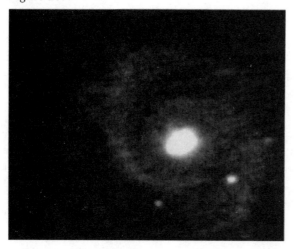

2. Save this image as a file called M5lstr.tif so your instructor can judge your progress.

Sharpening and Averaging over Discrepant Pixels (Noise Removal)

Here, although the contrast has been improved by stretching, there are many pixels in the image which are much different than their surroundings, giving the image a noisy appearance. These pixels are ones which have been modified by cosmic rays, random electrical noise, and other effects not related to the real ccd illumination. These image defects can be eliminated by replacing discrepant pixels with an average of their surrounding pixels (sometimes called noise removal or despeckling). Here, we must believe that any real image feature must be as large as the star images, which are several pixels in diameter. After processing in this way, the image is still not sharp-looking, where star images are smaller. To improve the sharpness of an image, we apply an image filter called sharpening. In this process, differences in brightness between adjacent pixels are accentuated. Figure 25.4, after noise removal, sharpening, and noise removal once more, becomes our final image, Figure 25.5 below.

Figure 25.5

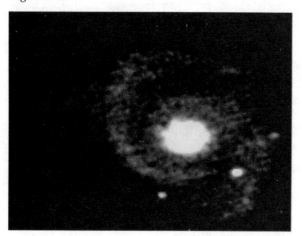

3. After performing these functions, save your image as the file M51final.tif.

Using False Color

As you can see in Figure 25.5, stretching an image will always lose the information in the region of the histogram outside of the range you have selected to display. Here, all the detail in the central part of the galaxy has been "overexposed," so that the supernova's existence can no longer even be inferred from the processed image. It is possible to preserve the information in all parts of the image by using "false colors." That is, for a black and white image (as most astronomical images are) let the image color stand for its brightness. Then, a wide range of brightnesses can be simultaneously viewed. Since this book is not reproduced in color, we can't show you such an image here, but you can get a flavor for the effect from Figure 25.6, where a false color version of Figure 25.5 is displayed in grayscale.

Figure 25.6

Sharpening Lunar or Planetary Images

Sometimes, an image is somewhat fuzzy, due to poor seeing or a slight error in focus. Image filters can be used to sharpen such an image up. These filters accentuate the difference between every pixel and its neighbors all around it or those above, below, left and right. In Figure 25.7, an image of the lunar crater Eratosthenes is shown, just as it was recorded on a CCD. As it stands, this image shows a great deal of detail. However, by using only image sharpening techniques, it was converted into Figure 25.8.

Much more detail is now displayed. One can even see fine ridges and elevation differences in the surrounding mare surface, and the edge of the great crater Copernicus is starting to become visible. However, a sharpened image will always contain more noise than the original. In this case, the level of noise is not objectionable, but in other images, such as

Figure 25.7

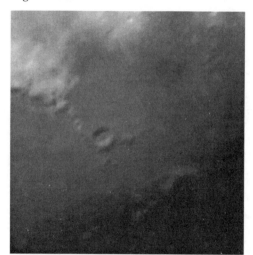

Figure 25.8

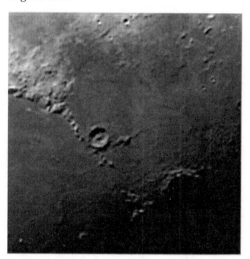

M51 we discussed earlier, further sharpening of the image is not recommended. Also notice that there is an image defect near the lower left corner limb of the image. Even excellent pictures are not perfect! Image sharpening works especially well on images of our moon and of planets.

4. You can achieve the above results in the following way: Load the image file moonfile.tif into your image processing software. Choose the filter image menu, and from that menu, choose sharpen. Save your result in a file called eratoshp.tif so your instructor can see it.

Other Image Processing Techniques

There are many other image processing techniques used on astronomical images. One, called maximum entropy, sometimes gives good results on images of marginal quality. In order to use it, a star image must also be in the image area, and the technique uses the size of the star image to attempt to maximize the image resolution. Other techniques worthy of mention are averaging, unsharp masking, smoothing, histogram equalization, and gamma correction. We will not attempt to describe these techniques here. But, there is nothing to keep you from trying them out. Experiment on the other images included with your software package. The guideline is, if a technique improves the image quality, use it, and if it doesn't, don't. Usually, you can tell if a technique has worked the desired magic of bringing out real details not evident in the original image.

Name:

Exercise Twenty-Six: Outdoor

CCD Photography at the Telescope

I. Astronomical CCD Photography

In tonight's lab you will be using a clock-driven telescope to photograph an astronomical object, using a CCD camera. You will also do preprocessing of the CCD image you have taken. For this lab you will work in groups of two: One will point and focus the telescope for the CCD image and one will take the exposure and do the preprocessing and storage of the image.

II. First Principles: Pixel Size and Resolution

It is important in doing CCD photography that you determine the scale factor of the image at the focal point. Since CCD chips are typically small in size, each pixel covers only a very small part of the focal plane. For example, a chip might only measure 2 mm on one side, but contain 192 pixels in that direction. In this case, each pixel is only about .01 mm across. Assuming that the telescope used to make the exposure has an effective focal length of 2 m, that .01 mm corresponds to an angle of only 1". If the resolution of the telescope and the atmospheric seeing are 1" or less, a star image will be less than one pixel in size. Thus, a perfectly focused image will put each star image primarily in only one image pixel.

If, on the other hand, the telescope resolving power or the atmospheric seeing is 1" or greater, even a perfectly focused star image will always occupy four pixels or more.

1. Find the size of your CCD's maximum image dimension from the CCD documentation.

   Write the value here: _____ mm.

2. Find the number of pixels across your CCD's maximum image dimension from the CCD documentation.

   Write the value here: _____ pixels.

3. Divide your value for number 1 above by your value for number 2, and place the answer here:

   CCD image scale = _____ mm/pixel.

4. Find the focal length of your telescope in mm from the telescope documentation, or from a procedure such as given in Exercise Ten, section III, and write the value here:

   _____ mm.

5. Since there are 2×10^5 seconds of arc per radian, you can find the scale of your telescope at the focal plane by dividing this number by the focal length value in question 4 above. Write the answer here:

   Telescope scale _____ "/mm.

6. Now, multiply your answer in 5 above by your answer to 3 to find how many seconds of arc there are will be per pixel in your image, and write the answer here:

   _____ "/pixel.

   The best angular resolution atmospheric seeing will ordinarily allow is about 1". In your telescope, would a 1" star image be larger than one pixel?

7. Using the formula given in Exercise Ten, section II, and reproduced here, calculate the angular resolution of your telescope.

   Angular resolution = 4.5"/D (inches)

   Angular resolution = _____ ".

8. Is your angular resolution smaller or larger than one pixel? Based on your answer and on your answer to question 6. above, should a perfectly focused star image be larger or smaller than one pixel in your image?

Exercise Twenty-Six: CCD Photography at the Telescope

III. Setting Up Your CCD Equipment

First, make sure you can accurately point your telescope with its finder scope. To do this, point the telescope at a bright star. With a high-power eyepiece, center the star in the field of view with the clock drive running. Next, without changing the telescope pointing, look through the finder, and make sure the star image is also centered in the finder field of view. If it is not, adjust the finder scope so that the star is accurately centered. Now remove the eyepiece from the telescope, so you can insert the CCD into the eyepiece holder.

Astronomical CCD cameras use computers to acquire, store, and process the images. For your CCD camera, insert the camera in the eyepiece holder, make the required connections, and start the software, according to the instructions in the CCD user's manual. Often, CCDs are cooled by electronic (so-called Peltier) coolers. If your CCD is cooled, use the software provided with your CCD to turn on its cooler. It may take 15 minutes or more for the cooler to reach a stable temperature, so this should be one of the first steps you take before making astronomical exposures.

1. Focus the telescope. This is done by pointing the telescope at a star and making rapid-succession CCD images while adjusting the focus. Most CCD camera software has a short time-exposure rapid-fire focus option. If this is the first time you have focused the CCD–telescope combination, it is probably best to focus on a bright star that you have centered in the finder scope field of view. Because the focus on a CCD is critical, and the range of motion of a telescope focuser is usually large, you should start out at one end of the focuser range and make coarse adjustments toward the other end until you see a badly out-of-focus image, and then make successively finer adjustments until the image is sharp. If you have gone through the entire focus range without seeing any image, you have either made adjustments that were too coarse, or the star was not in the CCD field of view. Start over, making sure that the star is precisely centered in the finder, and making somewhat finer initial adjustments than previously. Do not be surprised or frustrated if achieving a good focus takes five or ten minutes. Once a focus is achieved, mark the focuser position with a pencil, if possible, so a rough focus can be easily achieved in the future. If your software allows, save an image with your final focus with the name "focusnamedate".

   a. Special techniques for focusing.

      i. Aperture masks. It may be useful to use a commercially made or self-fabricated aperture mask to help focus on a star image. Usually, these consist of opaque discs with two or three circular apertures cut out of them. See Figure 26.1. Each of these apertures lets light through from a different part of the telescope objective. In a well-focused image, light from all parts of the objective is focused at the same point. Thus, in an out-of-focus star image, the apertures will be seen as separate out-of-focus images, whereas when the telescope is properly focused, the separate images will blend into a single image. This technique will only produce a perfectly

Exercise Twenty-Six: CCD Photography at the Telescope

Figure 26.1
An aperture mask for focusing. It should cover the entire telescope objective.

focused image if each aperture in the mask individually has an angular resolution better than the atmospheric resolution (about 1" of arc). If the objective is not large enough to have mask apertures over about 4.5 inches in diameter, the focus achieved will only be approximate. However, each out-of-focus star image appears double or triple in the CCD image, so it is easy to identify what is a real star image and what is not. Remember to remove the mask before taking your "real" exposures.

ii. Single-pixel focusing. When the focus is approximately correct, and the seeing and telescope angular resolutions are both smaller than one pixel, one can use the apparent star brightness in a pixel to do the final focusing. Using extremely fine focus adjustments, try to make each star image as bright as possible in its pixel. This is tantamount to focusing as much light as possible inside a single pixel, which is the desired goal. Of course, if a star image will not fit into a pixel in the setup you have, you can use the apparent sharpness of the image as the final focus indicator. You calculated whether this should be the case for your particular telescope setup in question II.8 above.

IV. Making the Exposure

Now, you should be ready to make exposures with your CCD camera. Exposure times will vary, depending on the brightness of the object you wish to photograph, the precision of your tracking, the sky brightness, etc. Typical exposure times for planets are 0.1 to 5 seconds. For faint nebulae, star clusters, and galaxies, exposures as long as 5 minutes may be appropriate. Make a few trial exposures just to find the right exposure time. The brightest important part of your image should not saturate the CCD (the brightest part of the image should not look perfectly white) in your exposure. If your CCD has a maximum image value per pixel of 4096, you should choose your exposure such that the maximum pixel value is just a little above 2048. For exposures that include overexposed bright stars, expect some blooming of their images, which in CCD images will look like streaks extending mainly to one side of the star. If these are objectionable, limit your exposure time until they are not a problem. It may be possible, using digital techniques, to add several short exposures so that none of them have "blooming" star images. For exposures of several minutes or more, your telescope clock drive may not be perfect enough to produce untrailed star images. Again, limiting the length of the exposure will produce images with no trails. For example, the tracking error on a telescope might be a second of arc per minute of time. If the seeing disc of stars is 3 seconds of arc on the night the exposures are made, one should limit the exposures to 3 minutes or less. And, for very long exposures, the "dark current" may saturate your image. Adding several short exposures is the only solution to this problem.

1. Make the exposure. Some CCD cameras and software require you to make a "flat-field image" and/or a "dark image" before your real exposure. In any case, it is a good idea to make these exposures and save them with names like "flatobjectnamedate" and "darkobjectnamedate". See the CCD software manual for details. Correcting your images for "flat-field" or "dark" subtraction is called preprocessing. Now, center the desired object in the field of view, and make your first exposure. Save it using a file name such as "objectdatenumber" (ask your instructor if you have any questions). You should make several exposures of the same object and save them with names that differ only in the "number" part of the name. One possible strategy for making these exposures is to save each one only if it is at least as sharp or detailed as the last one you saved. If it doesn't measure up, don't save it, but take another exposure. After half a dozen or so, you probably have one that is about as good as you are going to get. For each exposure you want to save, make sure to "flat-field" or "dark" correct it, if your software allows, before it is saved. Your instructor will examine your saved images for grading purposes. Make exposures of as many assigned objects as time allows, and save them with unique file names.

V. Image Processing

See Exercise Twenty-Five for many techniques you can use to improve the quality of your astronomical images. If your CCD camera allows exposures through different color filters, you may combine several such exposures to produce full-color images of your astronomical objects. Finally, for unsaturated star images, it may be possible to do photometry, using the total added pixel values for each star and the techniques described in Exercise Eighteen. Good luck and have fun with your images.

Exercise Twenty-Six: CCD Photography at the Telescope

Name: _____

Exercise Twenty-Eight: Indoor

Radial Velocities and the Hubble Law

I. Radial Velocities from Spectral Line Measurements

The standard method for finding radial velocities in astronomy is to measure the wavelengths of spectral lines. Objects that are moving away from the observer will have their spectral lines Doppler-shifted toward longer wavelengths than would be measured if they had no radial velocity. This is called the red shift. If the object is moving toward the observer, the spectral lines will be shifted toward shorter wavelengths (a blue shift). For radial velocities much smaller than the speed of light, the formula used to find the radial velocity is

$$V_r/c = (\lambda - \lambda_0)/\lambda_0$$

where λ is the astronomically measured wavelength of a certain line of a specific element, λ_0 is the wavelength measured in the laboratory, V_r is the radial velocity, and c is the speed of light (3.00×10^5 km/sec). This formula is only valid for small radial velocities, however (for V_r/c less than about 0.1).

1. Calculate V_r/c for a star moving 30 km/sec away from us. Place your answer here:

   $V_r/c =$ _____

2. Is the formula valid for a velocity this great?

3. Would the formula be valid for a velocity 200 times this great?

A formula that is valid for any value of radial velocity is found from Einstein's theory of relativity to be

$$V_r/c = [(\lambda/\lambda_0)^2 - 1]/[(\lambda/\lambda_0)^2 + 1]$$

For five galaxies, the wavelengths of the H and K lines of ionized calcium (prominent in the spectra of all galaxies) are to be taken from Table 28.1.

4. For each of these galaxies, first calculate the red shift of the H line and derive a velocity from its red shift.

5. Then calculate the red shift from the K line, and derive a velocity.

6. Finally, average the two velocity measurements for each galaxy to find the radial velocities of these galaxies. Place your answers here:

Galaxy 1: $V_r =$ _____ km/sec

Galaxy 2: $V_r =$ _____ km/sec

Galaxy 3: $V_r =$ _____ km/sec

Galaxy 4: $V_r =$ _____ km/sec

Galaxy 5: $V_r =$ _____ km/sec

Exercise Twenty-Eight: Radial Velocities and the Hubble Law

7. Although these are only hypothetical galaxies, notice that all of their radial velocities are positive (indicating motion *away from* the observer). This is true of most galaxies. How might you interpret this fact?

II. Distances of Galaxies

The distances to galaxies are difficult to determine. Lacking standard distance indicators such as trigonometric parallaxes (they are too far away to be measured), brightness of individual stars (they are too far away to be seen individually), or even standard diameters (they come in all sizes), faraway galaxies pose a problem in distance determination. Various methods have been attempted, such as using the diameters or brightnesses of globular clusters or giant HII regions. One promising method uses the rotationally broadened widths of 21-cm HI radio lines as an indicator of the total optical luminosity. For the farthest galaxies, the brightnesses of the galaxies themselves must be used. It has been found, for instance, that the 10th brightest galaxy in rich clusters of galaxies has pretty much the same brightness from cluster to cluster. In this lab, we will assume that the galaxies for which we have measured the red shifts in Section I are all about the same absolute magnitude, $M = -22.0$ (27 magnitudes brighter than the sun).

1. How many times the solar luminosity does each of our galaxies have? (Remember that 1 magnitude is a factor of 2.5 and every 5 magnitudes is a factor of 100.) Place your answer here:

$L =$ _____ $L_\odot$

2. Knowing the absolute magnitude of our galaxies, we can use their apparent magnitudes, m, to derive their distances, using the relation

$m - M = 5 \log d - 5$

Figure 28.1 is a plot of this relationship between d and m for m up to 20th magnitude and for $M = -22.0$. Using this figure and m given in Table 28.1, find the distances of our galaxies, and place your answers here:

Galaxy 1: $d =$ _____ pc

Galaxy 2: $d =$ _____ pc

Exercise Twenty-Eight: Radial Velocities and the Hubble Law

Galaxy 3: $d =$ _____ pc

Galaxy 4: $d =$ _____ pc

Galaxy 5: $d =$ _____ pc

3. One parsec is 3.26 light years. How far away is the most distant of our galaxies in light years?

4. The light we observe from the most distant of our galaxies was emitted a number of years ago equal to its distance in light years. The solar system is about 4.6 billion years old, and life on Earth is about 3 billion years old. How does the light from this distant galaxy compare in age with life on Earth and the solar system?

III. The Hubble Law

One of the most amazing facts about observing the faraway galaxies is that the interpretation of their red shifts and distances leads us to a model for the origin and evolution of the universe of galaxies. Hubble and Slipher were the first to publish a plot of the radial velocities of galaxies versus their distances from us. The interesting thing is that there is any relationship at all between these seemingly physically unrelated quantities.

1. Using the graph paper provided, make a Hubble plot of the radial velocities you found for our five galaxies versus their distances. Convenient units to use for the axes are 10^3 km/sec for radial velocity and 10^8 pc for distance. Draw the best straight line through the data points on your graph. Make sure your line passes through the origin.

What you have found, if your work so far is correct, is that there is a good straight-line relationship between galactic distance and galactic red shift. Interpreting this relation in terms of a real radial velocity dependent on distance, Hubble concluded that for such a relation to hold in a uniform (homogeneous) universe that looks the same in all directions (isotropic), the universe must be expanding uniformly in all directions. That is, the galaxies are all rushing away from each other in a uniform way. An equation called the Hubble Law relates the measured radial velocities to the distances in this way:

$V_r = H \times d$

Here, the constant H, which is the slope of the line on your Hubble plot, is called the Hubble constant. Astronomers now interpret the red shifts of faraway galaxies as being caused by the expansion of space itself, which carries the galaxies with it. This changes the formula for calculating the recessional velocity somewhat, but for the purposes of this exercise, the difference can and will be ignored.

2. Find H from your straight line and write your answer here:

   $H =$ _____ km/sec/pc

3. Astronomers usually use the units of km/sec/Mpc (kilometers per second per megaparsec) for H. Write H here in these units:

   $H =$ _____ km/sec/Mpc

If the universe is uniformly expanding at a constant rate, then there must have been a time when all of the galaxies were very close together. This can be seen in the following way. Suppose an arbitrary galaxy has been moving away from us at a constant radial velocity for a time t. It will have covered a distance $d = V_r \times t$ in that time. Substituting this into the Hubble Law, we find that $t = 1/H$. That is, the galaxy would have been here, at our location, the time $t = 1/H$ ago. Because we did this for an arbitrary galaxy, the result holds for any galaxy, and all galaxies must have been together this time ago.

4. In order to evaluate when this (the Big Bang) occurred, we must convert H into units of 1/sec. To do this, note that 1 pc $= 3.08 \times 10^{13}$ km. Write H here:

   $H =$ _____ /sec

5. Inverting H, find t (the age of the universe, assuming that the expansion has been at a constant speed), and place your answer here:

   $t = 1/H =$ _____ sec

Exercise Twenty-Eight: Radial Velocities and the Hubble Law

6. Finally, convert t to years, using the fact that 1 yr = 3.15×10^7 sec.

$t =$ _____ yr

Of course, in reality the expansion of the universe should be slowing because of the retarding effect of gravitation, so the galaxies have reached their present distances from us in a shorter time than if they had always traveled at their present, slower speeds. Thus, the age of the universe you have determined would be a maximum value, depending on how rapidly the expansion is slowing. It is a matter of contention whether the expansion is slowing rapidly enough for the universe to stop eventually and go into contraction. Recently, some data have indicated that the expansion may be accelerating. Clearly, further analysis is needed.

That the Big Bang actually occurred, however, is supported by observation of the highly red-shifted radiation from the early condensed state of the universe (the 3° blackbody background radiation) and by the agreement of the observed hydrogen and helium abundances with calculations of nuclear reactions occurring in the Big Bang. The maximum ages of stars, the change in the density of quasars with time in the universe, and other observational facts are consistent with the Big Bang.

Table 28.1

Apparent Magnitudes of Galaxies

Galaxy Number	Apparent Magnitude m	H Line λ	(λ/λ_0)	V_r/c	K Line λ	(λ/λ_0)	V_r/c	Average V_r/c
1	16.0	4231.2	_____	_____	4266.7	_____	_____	_____
2	9.4	3947.3	_____	_____	3982.8	_____	_____	_____
3	17.2	4483.5	_____	_____	4522.9	_____	_____	_____
4	15.1	4132.6	_____	_____	4168.1	_____	_____	_____
5	18.2	4834.4	_____	_____	4873.8	_____	_____	_____

Note: H Line of CaII λ_0 = 3933.7 Å
K Line of CaII λ_0 = 3968.5 Å
$c = 3.00 \times 10^5$ km/sec

Exercise Twenty-Eight: Radial Velocities and the Hubble Law

Figure 28.1

Relationship between *d* and *m* (*M* = –22.0).

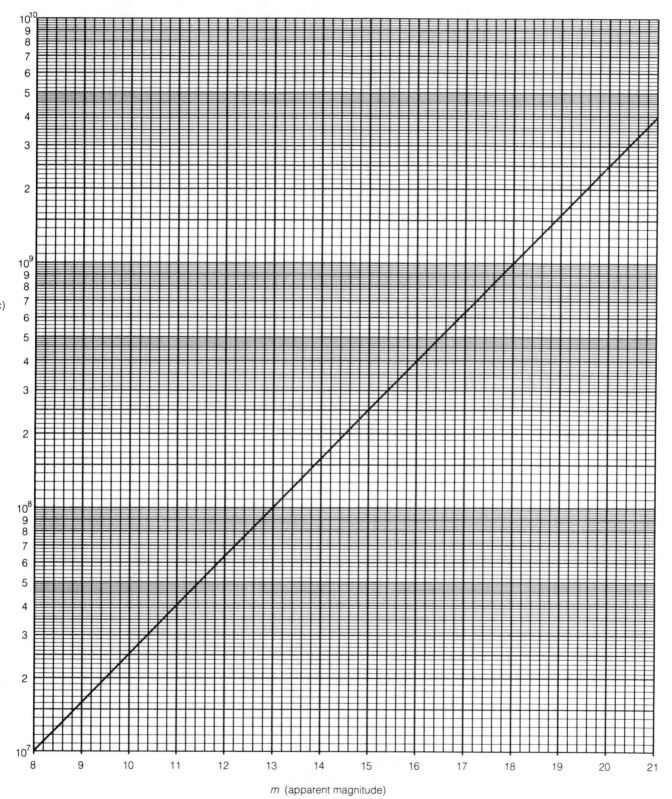

m (apparent magnitude)

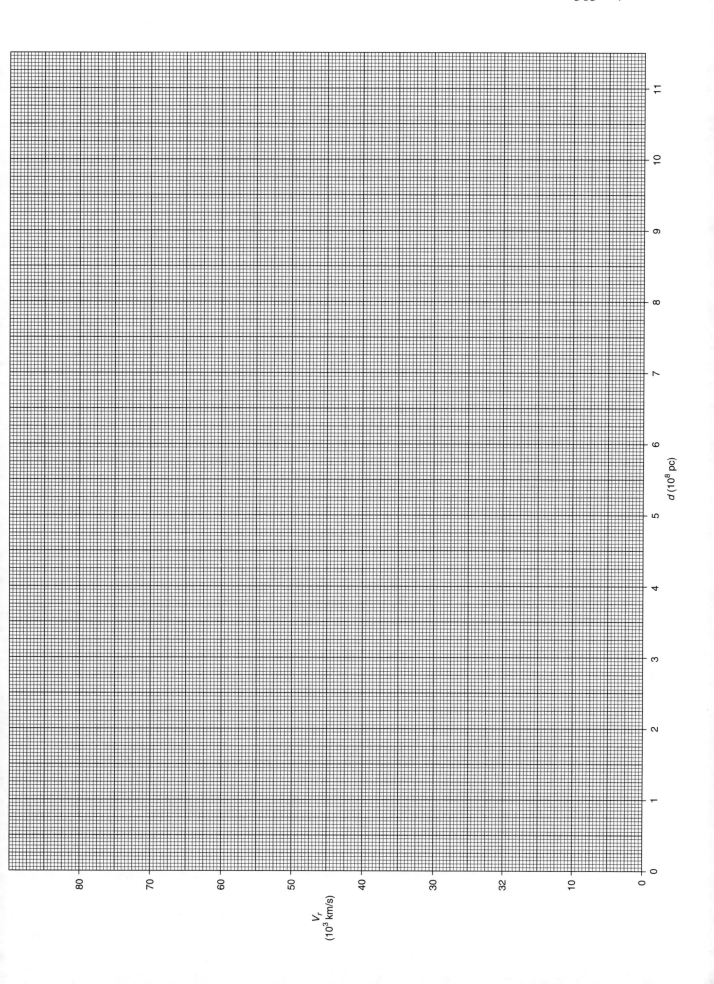

Name:

Appendix A

Fall Observing List

The following pages list a number of astronomical objects that you may be able to locate in the sky by the end of the term.

Throughout the term an observing period of about one hour's length may follow each lab. When you feel you are capable of finding one or more of these objects, you should attend one of these observing periods. Point out the object to the lab instructor, who may then check off the object on your list. By the end of the term all of the requirements should be checked off.

You should get started on this project as soon as possible, because some of the objects are visible only in the autumn and not at other times of the year. You may be allowed to use maps of the sky when pointing out these objects. However, advance preparation will avoid wasted time.

Throughout the term some of the lab exercises will help you to gain familiarity with the sky. Many of the objects on this list may be pointed out to you during the course of the labs. So if you pay attention during these labs, you may have little trouble completing the list.

Optional: While you are completing the observing list, you may wish to keep an observing notebook. Every time you observe one of the objects, you can make an appropriate entry in the notebook. For example, if the object is a constellation, you could sketch out the outlines formed by the brightest stars of the constellation. If the object is a lunar feature, a planet, or a star cluster, you could sketch its appearance as seen through a telescope or give a written description of its appearance. For each entry in the notebook you should also include the following information: place, instrument, weather conditions, date, and time. You may also wish to add descriptive notes concerning the method of observation. This notebook might then be turned in for extra credit (at the discretion of the instructor) at the end of the term.

I. Planets

Point out at least one of the following.

1. Mercury	5. Jupiter
2. Venus	6. Saturn
3. Mars	7. Uranus (you will need a telescope)
4. An asteroid (you will need a telescope)	8. Neptune (you will need a telescope)

II. Constellations

Find at least 10 of these.

1. Ursa Major	10. Sagitta	19. Triangulum
2. Ursa Minor	11. Delphinus	20. Perseus
3. Draco	12. Capricornus	21. Taurus
4. Cepheus	13. Aquarius	22. Orion
5. Cassiopeia	14. Pegasus	23. Lepus
6. Sagittarius	15. Andromeda	24. Auriga
7. Lyra	16. Pisces	25. Gemini
8. Aquila	17. Cetus	26. Canis Major
9. Cygnus	18. Aries	27. Canis Minor

III. Bright Stars

Find three of these.

1. Polaris (α UMi)
2. Vega (α Lyr)
3. Altair (α Aql)
4. Deneb (α Cygni)
5. Aldebaran (α Tau)
6. Capella (α Aur)
7. Betelgeuse (α Ori)
8. Rigel (β Ori)
9. Castor (α Gem)
10. Pollux (β Gem)

IV. Binary Stars

Find at least two.

1. Albireo (β Cygni)
2. Mizar and Alcor
3. Epsilon (ε) Lyrae
4. Almach (γ Andromedae)
5. Eta (η) Cassiopeiae
6. Alpha (α) Piscium

V. Open Clusters

Find one of these with a telescope.

1. Pleiades
2. Double cluster in Perseus (h and χ Persei)
3. M35 in Gemini
4. α Persei
5. M41 in Canis Majoris
6. Hyades

VI. Globular Clusters

These are strictly optional.

1. M15 in Pegasus
2. M2 in Aquarius
3. M22 in Sagittarius
4. M30 in Capricornus
5. M79 in Lepus

Appendix A: Fall Observing List

VII. Nebulae

Find at least one of these with a telescope.

1. M57: Ring Nebula in Lyra
2. M42 and M43: Great Nebula in Orion
3. M1: Crab Nebula in Taurus

VIII. Galaxies

Find one of these with a telescope.

1. M31: Andromeda Galaxy
2. M81 and M82 in Ursa Major
3. M33: Spiral galaxy in Triangulum

IX. The Moon

Find two of these lunar features with a telescope.

1. Copernicus
2. Plato
3. Tycho
4. Archimedes
5. Kepler
6. Mare Serenitatis
7. Mare Tranquillitatis
8. Mare Crisium
9. Mare Imbrium
10. Mare Nubium
11. Apennine Mountains

Name: _____

Appendix B: Field Trip

Finding List for October Celestial Objects

Table B.1 lists spectacular objects that should be seen easily in the evening on an early October night. They are listed in groups of more or less the same type of object. The order of listing within a group is the order in which the objects will set, so the first object in each group should be observed first. Any method (such as star charts or setting circles) may be used to find these objects. Those best seen with low powers are especially suitable for viewing through binoculars. Good luck and have fun!

Table B.1

October Celestial Objects

Name of Object	Constellation	Right Ascension (2000.0)	Declination (2000.0)	Remarks
		h m	(°)	
Open Clusters				
M7	Scorpius	17 54	−34.8	Use low power
M11	Scutum	18 51	−06.3	Medium power
h Persei	Perseus	2 20	+57.1	May get both h and χ
χ Persei	Perseus	2 23	+57.1	Persei in at lowest power
Globular Clusters				
M22	Sagittarius	18 36	−23.9	Best in low power
M13	Hercules	16 42	+36.5	Best in low power
M2	Aquarius	21 33	−00.8	Best in low power
Double Stars				
ε Lyrae	Lyra	18 44	+39.7	Quadruple star
Albireo or β Cygni	Cygnus	19 31	+28.0	Blue and gold Double; use high power
δ Cephei	Cepheus	22 29	+58.4	Orange and blue (brighter is variable)
Planetary and Diffuse Nebulae				
M8: Lagoon Nebula	Sagittarius	18 04	−24.4	Medium power OK on this diffuse nebula
M27: Dumbbell	Vulpecula Nebula	20 00	−22.7	Use low power on this nebula
M57: Ring Nebula	Lyra	18 54	+33.0	Best in high powers and is a planetary
Galaxies				
M31: Great Nebula	Andromeda in Andromeda	00 43	+41.3	Magnificent in a low-power eyepiece
M81	Ursa Major	9 56	+69.1	M81 is brighter than M82 and is a spiral
M82	Ursa Major	9 56	+69.7	M82 is a colliding galaxy and radio source

Appendix B: Finding List for October Celestial Objects

Name: _____

Appendix C

Spring Observing List

The following pages list a number of astronomical objects that you may be able to locate in the sky by the end of the term.

Throughout the term an observing period of about one hour's length may follow each lab. When you feel you are capable of finding one or more of these objects, you should attend one of these observing periods. Point out the object to the lab instructor, who may then check off the object on your list. By the end of the term all of the requirements should be checked off.

You should get started on this project as soon as possible, because some of the objects are visible only in the spring and not at other times of the year. You should be allowed to use maps of the sky when pointing out these objects. However, advance preparation will avoid wasted time.

I. Constellations

Find 10 of these.

1. Orion
2. Canis Major
3. Lepus
4. Canis Minor
5. Gemini
6. Auriga
7. Taurus
8. Cancer
9. Leo
10. Hydra
11. Corvus
12. Virgo
13. Boötes
14. Hercules
15. Corona Borealis
16. Ursa Major
17. Ursa Minor
18. Serpens
19. Libra
20. Ophiuchus
21. Draco

II. Bright Stars

Find five of these.

1. Betelgeuse
2. Rigel
3. Aldebaran
4. Procyon
5. Sirius
6. Castor
7. Pollux
8. Capella
9. Regulus
10. Arcturus
11. Spica
12. Denebola
13. Antares
14. Polaris
15. Zubeneschamali
16. Zubenelgenubi

III. Double Stars

Find two of these.

1. γ Leo
2. Castor
3. Alcor and Mizar in Ursa Major
4. γ Virgo

IV. Objects to Find with a Telescope

Find at least five of these.

1. M42: The Orion Nebula
2. M35: An open cluster in Gemini
3. M44: The Beehive cluster in Cancer
4. M67: An open cluster in Cancer
5. M81 and M82: Galaxies in Ursa Major
6. M36, M37, and M38: Open clusters in Auriga
7. M3: Globular cluster in Canes Venatici
8. M53: Globular cluster in Coma Berenices
9. M51, M94, and M106: Galaxies in Canes Venatici
10. M13 and M92: Globular clusters in Hercules
11. M5: Globular cluster in Serpens
12. M10 and M12: Globular clusters in Ophiuchus
13. M64: Spiral galaxy in Coma Berenices

Name: _____

Appendix D: Field Trip

Field Observing in March

I. Constellations

For all objects, see the finding charts: Figures D.1, D.2, D.3, D.4, and D.5.

Bright	*Dim*
Auriga	Eridanus
Taurus	Lepus
Orion	Monoceros
Gemini	Columba
Canis Major	Puppis
Canis Minor	Hydra
Leo	Cancer
Ursa Major	Leo Minor
	Canes Venatici

II. Binoculars

1. Look at the Pleiades in Taurus. How many stars can you see (approximately)?

2. Observe Praesepe (M44) in Cancer. This was one of Galileo's first discoveries. How many fuzzy patches (clusters or nebulae) can you see in the pentagon of Auriga?

3. Observe some of the double stars listed. From the magnitudes given, estimate the faintest star observable.

4. Hunt for some of the nebulae and clusters listed. Be sure to include M42 in Orion.

Double Stars for Binoculars

Here and in following lists, magnitudes are given in parentheses, separation afterward. Describe those you find, or have instructor check. Brace arms to hold binoculars steady. Test double for binoculars: ι Cancri (4, 7) 31".

γ Leporis (3.6, 6.4) 95"
θ² Orionis (5.4, 6.8) 52"
τ Tauri (4.5, 7.0) 63"
θ Tauri (3, 4) 5' 35"
σ Tauri (5, 5) 4' 40"
14 Canis Minoris (5.5, 7.8) 76", 112" (Triple. Draw the triangle as you see it.)
ζ Geminorum (4, 7) 94"
λ Aurigae (5.2, 8.7) 104"
ζ Ursa Majoris (Mizar) and Alcor (2, 5) 11' 47"

Clusters and Nebulae for Binoculars

M36, M37, and M38 in Auriga

M42 in Orion (Great Nebula)

M44 in Cancer (Praesepe)

M45 in Taurus (Pleiades) and the Hyades

M41 in Canis Major (An open cluster)

M35 in Gemini

M81 and M82 in Ursa Major (Not on finding chart; ask instructor)

III. Small Telescopes

Double Stars for Telescopes 3" or Larger

Use 100× if closer than about 10".

θ' Orionis (6.7, 7.5, 8) 9" (This is closest; center of Orion Nebula: excellent.)

σ Orionis (4, 10) 11", (7, 7.5) 13" (Quadruple.)

ι Orionis (3.2, 7.4) 11"

ρ Orionis (4.7, 8.5) 7"

λ Orionis (0.1, 6.7) 9.4" (Rigel.)

23 Orionis (5, 7) 32"

α Leporis (4, 9.5) 35"

β Monocerotis (4.7, 5.2, 5.6) 7", 10" (Beautiful triple.)

α Geminorum (2.0, 2.8) 5" (One of the best.)

δ Geminorum (3.2, 8.2) 7"

20 Geminorum (6.0, 6.9) 20"

38 Geminorum (5.4, 7.7) 7"

κ Geminorum (4.0, 8.5) 7"

ω Aurigae (5, 8) 6"

26 Aurigae (6, 8) 12"

14 Aurigae (5, 11, 7) 12", 14"(Triple.)

41 Aurigae (6.1, 6.8) 8"

ξ Tauri (5.5, 8) 20"

Σ 559 Tauri (7.0, 7.1) 3" (Use highest power on 3-inch.)

118 Tauri (5.8, 6.6) 4.8"

Appendix D: Field Observing in March

η' Canis Majoris (5.8, 7.9) 17"

ι Cancri (4, 7) 31"

κ Puppis (4, 5) 10"

γ Leonis (2.6, 3.8) 4" (One of the best; favorite of the nineteenth-century astronomer Struve.)

α Canes Venaticorum (2.9, 5.7) 20" (Very beautiful.)

ζ Ursa Majoris (2.4, 4.0) 14" (Excellent.)

Σ 1495 Ursa Majoris (6.0, 8.3) 34"

Clusters and Nebulae for the Telescope

All those listed for binoculars will be the best. Others listed here.

M67 in Cancer

M94 in Canes Venatici

M79 in Lepus

M3 in Canes Venatici (Probably not visible before 10:00.)

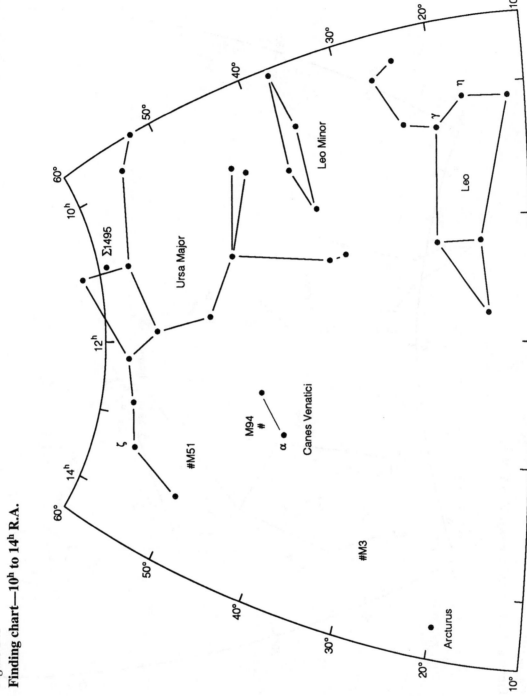

Figure D.1
Finding chart—10ʰ to 14ʰ R.A.

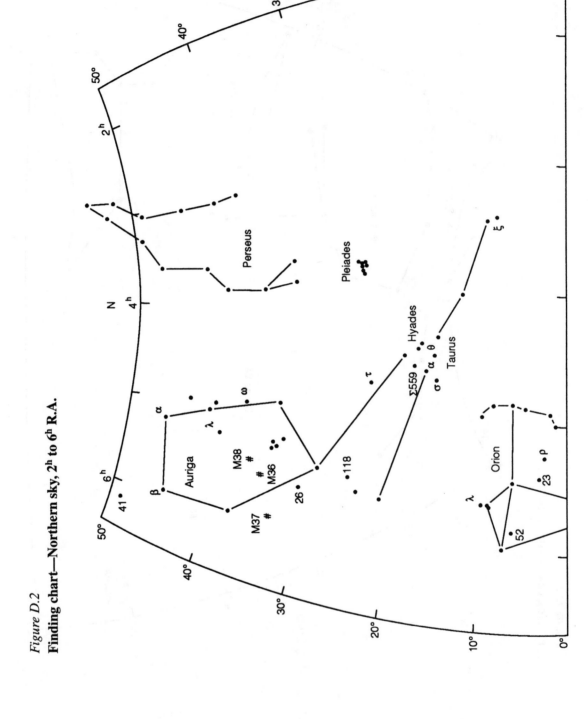

Figure D.2
Finding chart—Northern sky, 2ʰ to 6ʰ R.A.

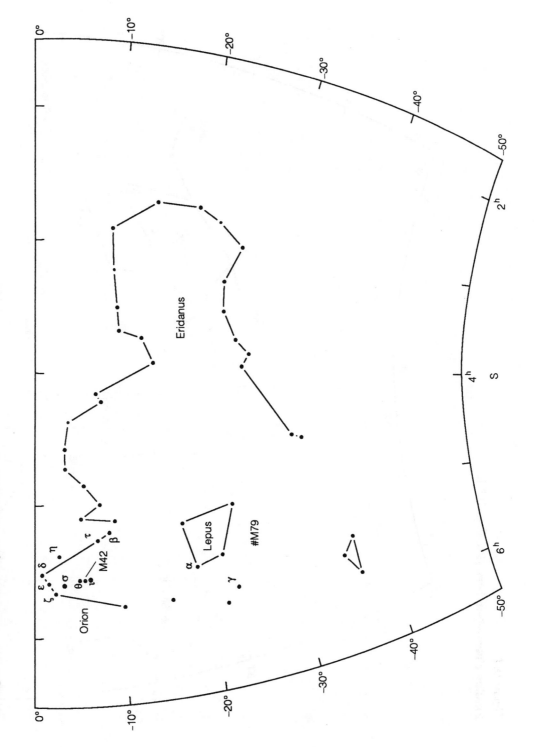

Figure D.3
Finding chart—Southern sky, 2ʰ to 6ʰ R.A.

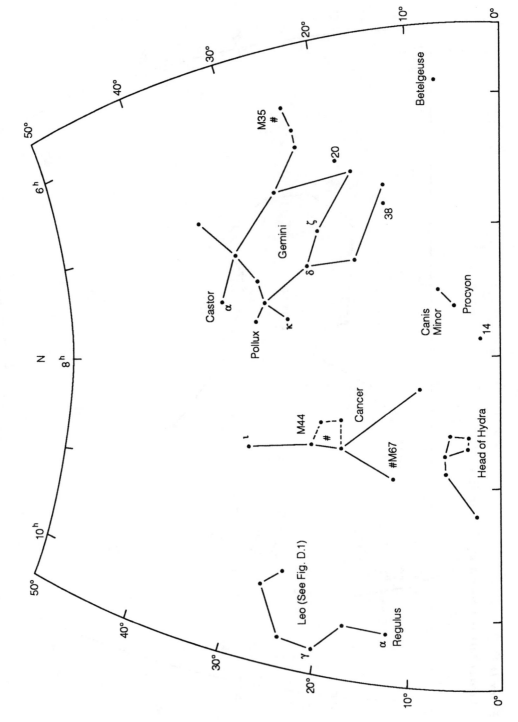

Figure D.4
Finding chart—Northern sky, 6ʰ to 10ʰ R.A.

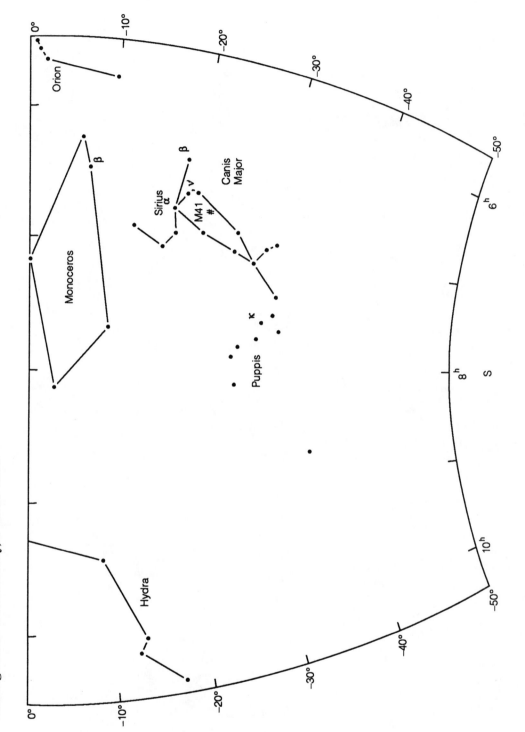

Figure D.5
Finding chart—Southern sky, 6ʰ to 10ʰ R.A.

Name: _____

Appendix E

Aligning a Telescope Axis

The following procedure will allow you to approximately align a telescope axis. It will require a calculation of the hour angle and declination of a bright southern star and the use of a carpenter's level, accurate to a fraction of a degree.

1. Set the elevation angle of the polar axis to equal your latitude. This may be done indoors in the following way:

    a. Using the level, make sure the base of your mounting is level.

    b. Point the telescope to the zenith (do this by laying the level across the mouth of the telescope and moving the telescope on its mounting until the mouth is perfectly level in both directions).

    c. Read the telescope declination circle. If it does not read your latitude, adjust the angle of elevation of the polar axis (see the instructions for your mounting on how to do this) so that it will read your latitude when the telescope is pointing to the zenith. For example: suppose that when the telescope is pointing to the zenith, the declination circle reads 35 degrees, whereas your latitude is 38.7 degrees. This means the polar axis elevation should be increased by 3.7 degrees. After adjustment, repeat steps a through c to make sure the adjustment is correct.

    d. Once properly adjusted, this polar axis elevation should remain the same as long as you observe in the same location (or within a few miles north or south).

2. At the start of observing, align the polar axis to true north. This may be done in the following way:

   a. With the telescope set up in the observing location, make sure the mounting is level and point the telescope to the zenith in the same way it was done in step 1 (using a level). The declination circle should read your latitude. If it doesn't, repeat step 1.

   b. With the drive motor turned off, adjust the right ascension slip-ring setting circle to 0 hours, or the hour angle circle to 0 hours, if so equipped.

   c. Find the hour angle and declination of a bright star in the southern sky. Depending on the time of year, good candidates might be Altair, Spica, Sirius, Antares, Procyon or another bright star (or planet).

   d. Turn the telescope on its two axes so that the setting circles read an hour angle equal to the hour angle of the bright star, and a declination equal to the declination of the bright star. If you have a slip-ring type right ascension circle, you may have to add 24 hours to the hour angle of the star. Unless you were extremely lucky, the telescope will not be pointing at the star in the sky.

   e. Physically rotate the telescope mounting on its supports until the bright star is in the field of view of the finder, and then centered in the telescope field of view. To do this, rotate it around its base (if possible) or rotate the tripod to adjust its alignment with true north. Make sure it remains level. If it has not taken more than a minute or so to do the last adjustment, no further adjustment of the right ascension or declination setting should be necessary to do this. At the end of this procedure, the polar axis of the telescope should now be pointing precisely toward true north. Make sure the hour angle is still correct. If not, repeat this step.

   f. Mark the position of the tripod legs on the surface, or of the telescope base on its support, so that true north may be found again at a later time without going through this whole procedure, if possible.

   g. If you have a slip-ring type of right ascension circle, now slip the ring to read the right ascension of the star. You may now turn on the clock drive motor.

   Your mounting and setting circles are now ready for use. Test them out by pointing the telescope to another star. Happy observing!

Appendix E: Aligning a Telescope Axis